THE FUNDAMENTAL INTERACTION

Geometrical Trends

THE FUNDAMENTAL INTERACTION

Geometrical Trends

Edited by

Joachim Debrus
Bad Honnef Physics Center
Bad Honnef, Federal Republic of Germany

and

Allen C. Hirshfeld
Physics Department
University of Dortmund
Dortmund, Federal Republic of Germany

PLENUM PRESS • NEW YORK AND LONDON

Library of Congress Cataloging in Publication Data

The Fundamental interaction: geometrical trends / edited by Joachim Debrus and Allen C. Hirshfeld.
p. cm.
"Proceedings of a meeting on the Fundamental interaction—geometrical trends, held February 16–20, 1987, at the Physics Center in Bad Honnef, Federal Republic of Germany"—T.p. verso.
Includes bibliographical references and index.
ISBN-13: 978-1-4615-9524-3 e-ISBN-13: 978-1-4615-9522-9
DOI: 10.1007/978-1-4615-9522-9
1. Field theory (Physics)—Congresses. 2. Gauge fields (Physics)—Congresses. 3. Particles (Nuclear physics)—Mathematics—Congresses. 4. Mathematical physics—Congresses. I. Debrus, Joachim. II. Hirshfeld, Allen C.
QC793.3.F5F86 1988 88-17661
530.1'4—dc19 CIP

Proceedings of a meeting on the Fundamental Interaction:
Geometrical Trends, held February 16–20, 1987, at the
Physics Center in Bad Honnef, Federal Republic of Germany

Softcover reprint of the hardcover 1st edition 1988
A Division of Plenum Publishing Corporation
233 Spring Street, New York, N.Y. 10013

PREFACE

A meeting was held at the Physics Centre in Bad Honnef from Feb. 16-20, 1987 on the subject "The Fundamental Interaction: Geometrical Trends". This meeting was in the series of Physics Schools organized by the German Physical Society. The participants were mainly younger scientists and graduate students: physicists, mathematicians and astronomers from the Federal Republic of Germany; there were also participants from Austria, the Netherlands and Switzerland. The purpose of the meeting was to introduce the participants to modern methods of mathematics and field theory which are increasingly being used in current elementary particle research. An outstanding feature of the school was the fact that each lecturer really made an effort to present his material at an introductory level, which could be followed by people with the usual degree level qualifications and which, nevertheless, then led up to the level of contemporary specialized literature. We hope that the published volume will make these lectures, which taken together give a unified overview of the recent developments leading up to the current world-view in modern theoretical physics, available to a wider audience.

We wish to thank the lecturers, who besides the formal lectures devoted their time to discussions with the participants throughout the duration of the School. We are also grateful to the staff of the Physics Centre for their organizational assistance, and to the participants, whose active interest made the meeting a success. The entire manuscript was efficiently typed by Manfred Krafczyk at the University of Dortmund. We are happy to acknowledge the generous support of the Volkswagen Foundation, which made all this possible.

J. Debrus
Physics Centre
Bad Honnef

A. C. Hirshfeld
Physics Department
University of Dortmund

CONTENTS

INTRODUCTION

After a period of relative stagnation, theoretical physics has in recent years entered a period of exciting growth: with new and unconventional ideas, and using a greatly increased mathematical arsenal, various groups are vigorously striving to achieve a deeper and more unified framework for understanding the world. In particular, Einstein's vision of a world formula, or fundamental law encompassing all physical processes, is again achieving prominence.

This fortunate breakthrough has its roots in two quite disparate - even in a certain sense opposing - directions of research: On the one hand we have an evergrowing wealth of empirical data, from the realm of the very large to the realm of the very small, from radio telescopes for investigating the furthest reaches of outer space to gigantic accelerators for probing the most minute constituents of matter. On the other hand, the introduction of new mathematical techniques, from the disciplines of geometry and algebra, some the fruits of contemporary mathematical research, has proved to be of fundamental importance for the erection of the conceptional framework in which the new physical theories are set.

It was this second - more mathematically oriented - aspect of the modern developments which formed the scientific context of the Bad-Honnef meeting. It is perhaps worth stressing at this point the important role (often underemphasized in introductory courses and textbooks) played by formal mathematical considerations in the foundations of all important physical theories. The compelling mathematical elegance of the fundamental laws of nature, as viewed from a modern vantage point, cannot be mistaken: think only of the symplectic geometry of Hamiltonian mechanics, the differential forms of thermodynamics, the two basis-free derivative operators of Maxwell's equations or the curvature scalar which is the Lagrangian for gravitation, to name only a few.

In this connection it is tempting to conclude that "understanding" a body of empirical facts in the physical sciences consists of reducing it to a natural and transparent mathematical structure. In some cases the appropriate structure already existed in mathematics, in an abstract form, when the need for it arose, in others the structures were first suggested by physical considerations, or, to put it more strongly, "discovered" by these means.

In modern developments we can see many examples of both kinds of these mutually beneficial interactions between mathematics and physics:

(i) The tremendous body of information on hadronic processes, over an exceedingly large energy range - from nuclear properties through middle-energy physics to the domain of contemporary elementary particle physics - can be reduced through Quantum Chromodynamics (QCD) to the geometrical gauge principle introduced by Weyl in 1929 and extended to non-Abelian groups by Yang and Mills in 1954. In the other direction, through the work of Atiyah and others, the concepts of non-Abelian gauge theories entered the domain of pure mathematics (here topology) and found important applications, for instance in Donaldson's proof of the existence of a second differentiable structure in $\mathbb{R}^4$.

(ii) The purely mathematical concept of a superalgebra plays a crucial role in contemporary theoretical physics; it was introduced by Wess and Zumino, who were led by physical considerations to generalize the invariance properties of relativistic quantum fields. This concept was immediately siezed on by several prominent mathematicians. The so-called Kac-Moody algebras, which emerged in investigations of the quantization of two-dimensional field theories, have undergone a similar fate.

(iii) One of the most fundamental insights of present day physics is the recognition that *all* the fundamental interactions- that is, besides QCD for the strong interaction, the Weinberg-Salam theory of the electro-weak interaction and, in a certain sense, Einstein's theory of gravitation - are governed by the gauge principle, and differ only in the choice of the gauge group. It was just this idea of the universality of interactions which led Weyl to his first theory involving a gauge principle, the 1919 theory for transporting lengths. This concept has been rehabilitated in modern physics, in a sense, within the framework of scale invariance.

These latter considerations lead naturally to the idea of an underlying common theory which will encompass the different theories discussed above and will yield the various interactions in different limits. From this point of view it is evident that the theory of gravitation must play a central role, and can be taken as a first approximation to the fundamental Ansatz. The first step in this direction was taken already in the 1920's by Klein and Kaluza: their aim was the unification of gravitation with Maxwell's electrodynamics. The second step was taken only many years later, when the extension of gravity to supergravity was motivated by the (unfortunately still unsolved) renormalization problem of quantum gravity. The third and, until now, last step is more far-reaching and generalizes the two previous attempts in a fundamental way. This is, of course, superstring theory. All at once, the three different structures associated with the weak, strong and gravitational interactions appear as different facets of an underlying reality. A well-founded hope for the resolution of the renormalization problem of quantum gravity seems at last to be at hand. It is believed that superstrings represent a finite theory, which avoids completely the necessity of the renormalization prescription considered by many to be unsatisfactory.

The ambitious superstring program is today still far from completion. Its immediate aims are less the detailed calculation of specific physical processes than the fundamental unsolved consistency problems of quantum theory and general relativity and their unification. In this respect there is a remarkable similarity to the consistency requirements between special relativity and Newtonian gravity, which first led Einstein to develop the general theory of relativity. Then too a (for that time) dramatic extension of the mathematical repertoire of physics was necessary, i.e. Riemannian geometry. Whereas the mathematical concepts had been developed by Gauss and Riemann and others many years before, decisive experimental verification, achieved in part through the use of radio telescopes, had to wait for nearly half a century.

In this general context it might be worthwhile to re-emphasize that in string theory gravity reassumes the central role in physics which was envisaged for it by Einstein, after a long period in which it was isolated from the mainstream. Einstein himself, despite many efforts in his later years, could not achieve this goal because he lacked the means to incorporate quantum theory in a fundamental way.

The contributions to this volume, which for the most part have the character of introductory courses and review articles, largely cover the general developments which could only be outlined above, and which were fundamental for the emergence of today's conceptional framework.

The first section deals with general methods of modern field theory. T. Schücker gives a modern presentation of the fundamental methods of differential geometry as applied to gauge theories. Such expositions are unfortunately still lacking in the current physics syllabus. N. Falck deals with the quantization of constrained systems, which is important for any quantum theoretical treatment of gauge theories, and which he applies in his second lecture to the question of the role of anomalies in the chiral Schwinger model. Finally, H. Papadopoulos discusses the decisive problem of compactification of higher spacetime dimensions. These far-reaching generalizations of the Kaluza-Klein idea could, regrettably, because of limitations of time and space, only be briefly sketched.

The second section of the volume is concerned with the problem of anomalies in quantum field theory. It was the study of this problem which led, for one thing, to an appreciation of the practical consequences of the renormalizability of a field theory, and for another ushered in the modern renaissance of string theory. H. Leutwyler elucidates the origin and significance of chiral and gravitational anomalies. N. Falck discusses the renewed interest in the quantization of the chiral Schwinger model, which may lead to a reassessment of the relation between anomalies and gauge invariance. M. Reuter's article deals with anomalies in different dimensions, the relationships between them and their relevance in various physical situations.

The third section of this volume deals with the so-called "supertheories". M. Sohnius' article gives an elementary introduction to the techniques of supersymmetry calculations, demonstrated for two-dimensional field theories. This case has the double advantage of being computationally simple and of playing a key role in the superstring theories. A. Hirshfeld first gives a general introduction to the subject of Lie superalgebras, a subject which is still lacking in the textbook literature. Even at the purely formal level, this topic represents a fascinating and impressive generalization of the conventional Lie algebras which are basic for the description of all hitherto studied symmetries in physics. He then goes on to discuss the relationship between gravitation and Yang-Mills gauge theories, a topic discussed by Schücker in the context of Cartan's formulation of gravitational physics. Here the group theoretical aspects are emphasized, and Einstein gravity appears as the limit of a gauge theory involving the anti-de-Sitter group as a local symmetry. With this groundwork supergravity appears as a straightforward supersymmetric extension of ordinary gravity theory. Finally, the article of H. Nicolai contains both a historical survey and a thorough discussion of some of the fundamental questions of string theory. Characteristic features connected with the quantization of bosonic strings are explained, such as the existence of a critical dimension and of tachyons. The treatment of the fermionic degrees of freedom, which when united with the bosonic ones yield superstrings, was discussed at the meeting by G. Münster from the University of Hamburg, but is covered in this published volume in Nicolai's article. The essential concepts of superstring theory are dealt with: the spinning string, the Neveu-Schwarz and Ramond models, the Gliozzi-Scherk-Olive projection operators which resolve the tachyon problem, open and closed superstrings. In summary a review is given of all presently known string theories in 10 dimensions.

Taken together, the lectures at this school, which were followed with great interest by about 60 young scientists, represent an important and necessary extension of the usual university syllabus. There is, in fact, a discrepancy between the subject matter of university graduate courses, and the prerequisites, especially in modern mathematics, necessary to follow and participate in the new developments in theoretical physics. This gap could be closed in the future only by a comprehensive reform of the whole traditional curriculum. This pressing question was discussed at the meeting in a lively evening discussion session (in the famous Bad-Honnef wine cellar). In particular, the necessary coordination of the instruction in physics and mathematics was pointed out to be a prerequisite for a better mutual understanding. Another important point was that the reformulation and generalization of mathematical methods and concepts, necessary on the one hand for the needs of current research, would on the other hand benefit the more elementary, largely classical fields of physics, where a substantial simplification and unification would result. Also under this more paedagogical aspect the Volkswagen-Foundation deserves our gratitude for its support of the present meeting.

K. Bleuler, M. Werner

Bonn, 26 March 1987

I. METHODS IN MODERN FIELD THEORY

DIFFERENTIAL FORMS AND GAUGE THEORIES

M. Göckeler and T. Schücker

Institut für Theoretische Physik der Universtät Heidelberg
Philosophenweg 16, 6900 Heidelberg, Fed. Rep. Germany

ABSTRACT

An introductory review is presented of differential forms and their use in physics, particularly with regard to gauge theories. The subjects covered are:

1. Vector fields
2. Differential forms
3. The wedge product
4. The exterior derivative
5. Integration
6. Vector valued differential forms
7. Frames
8 Metrics on a vector space
9. Metrics on an open subset of $\mathbb{R}^n$
10. The Hodge star
11. The coderivative
12. The Laplace operator
13. Summary of the mathematical part
14. Maxwell's equations
15. Gauge theories and the Yang-Mills action
16. The equivalence principle
17. The Einstein-Cartan equations
18. A farewell to ω

INTRODUCTION

The purpose of this lecture is a quick review of differential forms on open subsets of $\mathbb{R}^n$ and their use in formulating two well-established theories, Yang-Mills theory and general relativity. Of course, there is nothing new in this material and it may be found in many textbooks. The presentation will be elementary: Everything, except for the few theorems cited, can be verified by short, straightforward calculations, most of them constituting excellent exercises. The mathematical part contains nothing a physicist has not learnt during the first two years of his studies, only some notations may be unfamiliar.

1. VECTOR FIELDS

Let U be an open subset of $\mathbb{R}^n$. A vector field v on U is a differentiable family v(x) of vectors in $\mathbb{R}^n$ indexed by the points in U (for us differentiable always means infinitely many times differentiable). For example, U might be a lake and v might describe the currents on its surface. Note that the "velocity" vectors v(x) are not confined to lie in a subset of $\mathbb{R}^n$ as is the case for the points x.

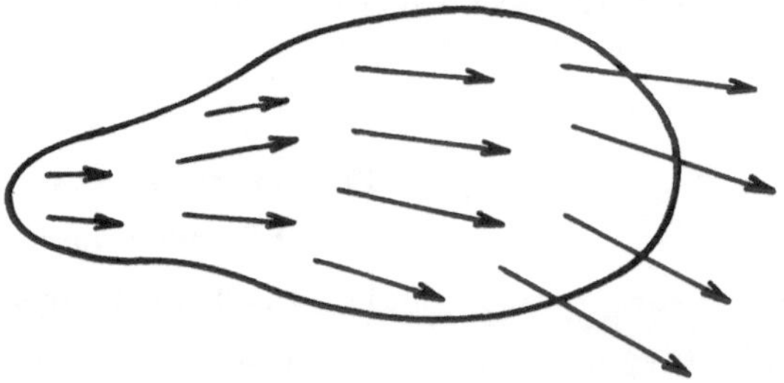

In Cartesian coordinates y^μ, $\mu = 1,2,\ldots,n$, any vector field may be decomposed as:

$$v = \sum_{\mu=1}^{n} v^\mu(x) \frac{\partial}{\partial y^\mu}, \tag{1.1}$$

where $\partial/\partial y^\mu$ are the vector fields with Cartesian components $(0,\ldots,0,\underset{\mu}{1},0,\ldots,0)$. For example, $\partial/\partial y^1$:

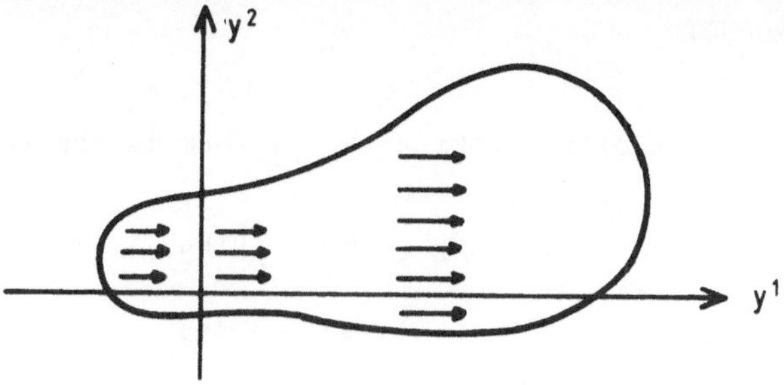

Note that here $\partial/\partial y^\mu$ is not a differential operator, but just a symbol. Its mnemo-technical utility comes from the definition in arbitrary coordinates x^μ :

$$\frac{\partial}{\partial x^\mu}(x) := \sum_\nu \frac{\partial y^\nu}{\partial x^\mu}(x) \frac{\partial}{\partial y^\nu}, \tag{1.2}$$

where $\partial y^\nu/\partial x^\mu$ is the Jacobian matrix of the (general) coordinate transformation. We shall consider explicitly the example of polar coordinates in chapter 7.

2. DIFFERENTIAL FORMS

By definition a (differential) p-form ϕ is a differentiable family of maps ϕ_x

$$\phi_x : \mathbb{R}^n \times \ldots \times \mathbb{R}^n \to \mathbb{R}$$
$$(v_1(x), \ldots, v_p(x)) \mapsto \phi_x(v_1(x), \ldots, v_p(x)).$$

Each map ϕ_x is required to be multilinear (with respect to the real numbers) and alternating, i.e.

$$\phi(\ldots, v_i, \ldots, v_j, \ldots) = -\phi(\ldots, v_j, \ldots, v_i, \ldots) . \tag{2.1}$$

For convenience we often suppress the point x. We denote by $\Lambda^p U$ the set of all p-forms on U. Note that if $p > n$ this set only contains the zero element. For p=0 we define $\Lambda^0 U$ to be the set of all (differentiable) functions from U into the real numbers.

3. THE WEDGE PRODUCT

The wedge product of a p-form with a q-form is the (p+q)-form defined by:

$$\wedge : \quad \Lambda^p U \times \Lambda^q U \to \Lambda^{p+q} U$$

$$(\phi, \psi) \mapsto \phi \wedge \psi$$

$$(\phi \wedge \psi)(v_1, \dots, v_{p+q}) := \frac{1}{p!q!} \sum_{\pi \in S_{p+q}} \operatorname{sign}\pi \; \phi(v_{\pi(1)}, \dots, v_{\pi(p)}) \cdot \psi(v_{\pi(p+1)}, \dots, v_{\pi(p+q)}) \tag{3.1}$$

where the sum is over all permutations of p+q objects and signπ is the sign of the permutation π.

The wedge product is bilinear, associative and graded commutative, i.e.

$$\phi \wedge \psi = (-1)^{pq} \psi \wedge \phi . \tag{3.2}$$

In any coordinate system x^μ a p-form may now be decomposed as

$$\phi = \sum_{\mu_1, \dots, \mu_p} \phi_{\mu_1 \dots \mu_p}(x) \, dx^{\mu_1} \wedge \dots \wedge dx^{\mu_p} , \tag{3.3}$$

where for each $\mu = 1,2,\dots,n$, dx^μ is the 1-form defined by

$$dx^\mu \left(\frac{\partial}{\partial x^\nu}\right) = \delta^\mu_\nu . \tag{3.4}$$

In tensor language a vector field v constitutes a contravariant tensor v^μ of degree (rank) one, while a p-form constitutes a completely antisymmetric covariant tensor $\phi_{\mu_1 \dots \mu_p}$ of degree p. The real number obtained by evaluating a p-form on p vector fields corresponds to the complete contraction and the wedge product corresponds to the antisymmetrized tensor product of antisymmetric covariant tensors.

A collection of vector spaces Λ^p, $p = 0,1,\dots,n$, together with a bilinear, associative, graded commutative product $\wedge$ is also called an exterior algebra, or Grassmann algebra.

4. THE EXTERIOR DERIVATIVE

We define the exterior derivative of a form using a coordinate system x^μ :

$$d: \Lambda^p U \rightarrow \Lambda^{p+1} U$$
$$\phi \mapsto d\phi$$

$$d\phi := \sum_{\mu_1,\ldots,\mu_p,\nu} \left[\frac{\partial}{\partial x^\nu} \phi_{\mu_1\ldots\mu_p}\right] dx^\nu \wedge dx^{\mu_1} \wedge\ldots\wedge dx^{\mu_p} . \qquad (4.1)$$

This definition does not depend on the choice of the coodinate system x^μ.

The exterior derivative is a linear first order differential operator. It obeys the Leibniz rule

$$d(\phi \wedge \psi) = (d\phi) \wedge \psi + (-1)^p \phi \wedge d\psi \qquad (4.2)$$

and the so-called co-boundary condition

$$d^2 = 0 . \qquad (4.3)$$

In tensor language the exterior derivative amounts to taking the gradient of an antisymmetric covariant tensor and then antisymmetrizing the covariant index of the gradient with the others. The co-boundary condition is just the statement that partial derivatives commute.

5. INTEGRATION

Let ϕ be a p-form and K a p-dimensional sufficiently regular piece of U parametrized by $x^1, x^2,\ldots,x^p$, for example a cube. Then we define the integral of ϕ over K:

$$\int_K \phi := \int_K \phi_{12\ldots p}\, dx^1\ldots dx^p , \qquad (5.1)$$

where the right-hand side is just the multiple Riemannian integral of the coefficient function of ϕ. The increasing order of the indices in the coefficient function $\phi_{12\ldots p}$ means that we suppose a fixed numbering of the coordinates of K, i.e. an orientation. The definition of the integral of a form does not depend on the choice of the coordinate system. This is assured by the theorem that under a change of coordinates the integrand in the Riemannian integral changes with the absolute value of the determinant of the Jacobian matrix.

Let us mention *Stoke's theorem*: Let ϕ be a (p-1)-form, K a p-dimensional piece of U, ∂K its properly oriented boundary. Then

$$\int_K d\phi = \int_{\partial K} \phi . \tag{5.2}$$

This theorem will be useful later on when deriving field equations from an action. Together with the Leibniz rule (4.2) it allows us to carry out partial integrations. Finally, we remark that the boundary of a boundary is empty,

$$\partial\partial K = 0 ,$$

which explains the term co-boundary condition for equation (4.3).

6. VECTOR VALUED DIFFERENTIAL FORMS

Let W be a finite dimensional real vector space. Since all operations introduced so far are linear, we can generalize the values of differential forms from the real numbers to vectors in W:

$$\phi_x : \mathbb{R}^n \times \ldots \times \mathbb{R}^n \rightarrow W .$$

We denote by $\Lambda^p(U,W)$ the set of p-forms on U with values in W. In later applications W will be a Lie algebra or a vector space carrying a linear representation of some symmetry group (in supersymmetric theories W is an abstract Grassmann algebra). With respect to a basis T_a, a=1,2,....,dim W, any element $w \in W$ can be written as

$$w = \sum_{a=1}^{\dim W} w^a T_a , \tag{6.1}$$

where the w^a are real numbers. Likewise any p-form ϕ with values in W can be written as

$$\phi = \sum_a \phi^a T_a , \tag{6.2}$$

where now the ϕ^a are real valued differential forms on U. Of course, in order to define a wedge product in this more general setting, W must have a multiplication law, i.e. W must be an algebra. For example, if W is a Lie algebra, we define the commutator of a p-form and a q-form, both with values in the Lie algebra, by

$$[\phi,\Psi](v_1,\dots,v_{p+q}) = \frac{1}{p!q!} \sum_{\pi\in S_{p+q}} \operatorname{sign}\pi\, [\phi(v_{\pi(1)},\dots,v_{\pi(p)}),\Psi(v_{\pi(p+1)},\dots,v_{\pi(p+q)})] \tag{6.3}$$

or, with respect to a basis T_a,

$$\phi = \sum_a \phi_a T^a , \tag{6.4}$$

$$\Psi = \sum_b \psi_b T^b , \tag{6.5}$$

$$[\phi,\Psi] = \sum_{a,b} \phi_a \wedge \psi_b\, [T^a,T^b] . \tag{6.6}$$

The commutator of forms is graded commutative:

$$[\phi,\Psi] = -(-1)^{pq}[\Psi,\phi] , \tag{6.7}$$

where one minus sign comes from the anticommutativity of the commutator of two Lie algebra elements and the others from equation (3.2).

7. FRAMES

A frame on an open subset U of $\mathbb{R}^n$ is a set of n vector fields $b_1,b_2,\dots,b_n$ such that in each point $x \in U$ the n vectors $b_1(x),\dots,b_n(x)$ are linearly independent. Other words used for frames are tetrads (for n=4), vielbein or n-bein, repére (mobile). If x^μ is a coordinate system, then
$\partial/\partial x^\mu$, $\mu = 1, 2, \dots, n$, is a frame. However, not every frame b_i can be derived from a coordinate system and we call a frame of the particular kind $\partial/\partial x^\mu$ holonomic. Later on we shall learn a recipe which tells us how to decide whether or not a given frame is holonomic.

Given two frames b_i and b'_i on U, we can always, at a given point x, expand one in terms of the other:

$$b'_i(x) = \sum_j (\gamma^{-1}(x))^j{}_i b_j(x) , \tag{7.1}$$

where $\gamma^{-1}(x)$ is an invertible n×n matrix:

$$\gamma^{-1}(x) \in GL_n . \tag{7.2}$$

Both frames depend differentiably on x and so does $\gamma^{-1}(x)$, i.e. γ^{-1} is a differentiable function from U into GL_n. The set of all such functions forms a group, where the multiplication is defined pointwise by the matrix product. We call this group the GL_n gauge group

$$^{U}GL_n = \{\gamma: U \to GL_n\} . \tag{7.3}$$

A dual frame (or simply frame, when there is no risk of confusion) is a set of n 1-forms $\beta^1, \beta^2, \ldots, \beta^n$, such that for every $x \in U$ $\beta^1(x), \beta^2(x), \ldots, \beta^n(x)$ are linearly independent. A frame is called holonomic if it is of the form dx^μ, where x^μ is a coordinate system.

Theorem: Let U be simply connected. Then the frame β^i is holonomic if and only if

$$d\beta^i = 0 \tag{7.4}$$

for $i = 1, 2, \ldots, n$.

A dual frame β^i is called dual to a frame b_i if

$$\beta^i(b_j) = \delta^i_j \tag{7.5}$$

for $i, j = 1, 2, \ldots, n$ and a frame is holonomic if and only if its dual frame is holonomic. If two frames b_i and b_i' are related by the gauge transformation γ^{-1}, Eq. (7.1), their corresponding dual frames are related by the inverse transposed gauge transformation:

$$\beta'^i = \sum_j \gamma^i{}_j \beta^j , \tag{7.6}$$

transposed because of the "wrong" order of the indices in Eq. (7.1). Our convention is that the first index of a matrix numerates the rows, the second index the columns, irrespective of whether the indices are upper or lower.

As an example let us consider three-dimensional polar coordinates, U is $\mathbb{R}^3$ without the x-z half plane:

$$U = \mathbb{R}^3 - \{(x,y,z),\ x \geq 0,\ y = 0\} . \tag{7.7}$$

Let b_i be the holonomic frame of Cartesian coordinates,

$$b_1 = \frac{\partial}{\partial x}, \quad b_2 = \frac{\partial}{\partial y}, \quad b_3 = \frac{\partial}{\partial z}, \tag{7.8}$$

and b_i' the holonomic frame of polar coordinates,

$$b_1' = \frac{\partial}{\partial r}, \quad b_2' = \frac{\partial}{\partial \varphi}, \quad b_3' = \frac{\partial}{\partial \theta}, \tag{7.9}$$

with

$$x = r \cos\varphi \sin\theta \tag{7.10}$$

$$y = r \sin\varphi \sin\varphi \tag{7.11}$$

$$z = r \cos\theta. \tag{7.12}$$

In order to calculate the gauge transformation γ relating the two frames we use the definition (1.2):

$$\frac{\partial}{\partial r} = \frac{\partial x}{\partial r}\frac{\partial}{\partial x} + \frac{\partial y}{\partial r}\frac{\partial}{\partial y} + \frac{\partial z}{\partial r}\frac{\partial}{\partial z} \tag{7.13}$$

and two similar identities; γ^{-1} is just the Jacobian matrix of equations (7.10-12):

$$\gamma^{-1} = \begin{bmatrix} \cos\varphi\sin\theta & -r\sin\varphi\sin\theta & r\cos\varphi\cos\theta \\ \sin\varphi\sin\theta & r\cos\varphi\sin\theta & r\sin\varphi\cos\theta \\ \cos\theta & 0 & -r\sin\theta \end{bmatrix} . \tag{7.14}$$

The corresponding holonomic dual frames are given by

$$\beta^1 = dx \qquad \beta^2 = dy \qquad \beta^3 = dz , \tag{7.15}$$

and

$$\beta'^1 = dr \qquad \beta'^2 = d\varphi \qquad \beta'^3 = d\theta . \tag{7.16}$$

Using Eq. (7.6) we then find

$$dx = \frac{\partial x}{\partial r} dr + \frac{\partial x}{\partial \varphi} d\varphi + \frac{\partial x}{\partial \theta} d\theta \tag{7.17}$$

and similar equations for dy and dz.

8. METRICS ON A VECTOR SPACE

Let V be an n-dimensional real vector space. A (pseudo-)metric (or scalar product) on V is a bilinear form

$$g : V \times V \to \mathbb{R}$$
$$(v,w) \mapsto g(v,w)$$

which is symmetric,

$$g(v,w) = g(w,v) \quad \text{for all } v,w \in V \tag{8.1}$$

and nondegenerate. The last requirement means that only the zero vector has vanishing scalar product with all vectors in V. If $b_1, b_2, \ldots, b_n$ is a basis of V, then, due to the bilinearity, the metric g is uniquely specified by the nxn matrix of the scalar products of the basis vectors:

$$g_{ij} := g(b_i, b_j) . \tag{8.2}$$

The symmetry and nondegeneracy of g imply that the matrix of g with respect to the basis is symmetric and nondegenerate:

$$g_{ij} = g_{ji} , \tag{8.3}$$

$$\det(g_{ij}) \neq 0 . \tag{8.4}$$

The matrix g'_{ij} of the metric g with respect to a different basis b'_i,

$$b'_i = \sum_i (\gamma^{-1})^j{}_i b_j \;, \tag{8.5}$$

is given by

$$g'_{ij} := g(b'_i, b'_j) = (\gamma^{-1T} g \gamma^{-1})_{ij} \;. \tag{8.6}$$

Note here that we use nxn matrices to describe a change of coordinates as well as a metric, two quite different mathematical objects.

The following two theorems of linear algebra are of fundamental importance for us.

Theorem (Gram & Schmidt): Any metric has an orthonormal basis e_i, i.e. a basis such that

$$g(e_i, e_j) = \eta_{ij} = \left[\begin{array}{c|c} \begin{matrix}1 & & \\ & \ddots & \\ & & 1\end{matrix} & \\ \hline & \begin{matrix}-1 & & \\ & \ddots & \\ & & -1\end{matrix} \end{array}\right] \begin{matrix} r \\ \\ s \end{matrix} \tag{8.7}$$

Theorem (Sylvester): The number r of plus signs and the number s of minus signs, r+s = n, does not depend on the choice of the orthonormal basis e_i.

From now on we shall reserve the letter e for an orthonormal basis. Of course, an orthonormal basis is not unique, for instance

$$e_1 = \begin{bmatrix}1\\0\end{bmatrix}, \qquad e_2 = \begin{bmatrix}0\\1\end{bmatrix} \tag{8.8}$$

and

$$e_1' = \frac{1}{\sqrt{2}}\begin{bmatrix}1\\1\end{bmatrix}, \quad e_2' = \frac{1}{\sqrt{2}}\begin{bmatrix}-1\\+1\end{bmatrix} \tag{8.9}$$

are both orthonormal for the Euclidian metric of $\mathbb{R}^2$. In general, given an orthonormal basis e_i, any other basis e'_i with

$$e'_i = \sum_j (\Lambda^{-1})^j{}_i\, e_j \;, \qquad \Lambda \in GL_n \tag{8.10}$$

is also orthonormal if and only if

$$\eta = \Lambda^{-1T}\, \eta\, \Lambda^{-1} \;. \tag{8.11}$$

The set of all Λ's satisfying this condition forms a subgroup of GL_n, the Lorentz group, denoted by O(r,s). It is of dimension $\frac{1}{2}n(n-1)$.

There are two ways to parametrize all possible metrics with given signature (r,s) on V.

a) Choose a fixed basis b_i of V. Then any metric is parametrized by the symmetric n×n matrix g_{ij} of scalar products, that is $\frac{1}{2}n(n+1)$ real numbers.

b) Given any metric, choose an orthonormal basis e_i. This basis characterizes the metric as well. With respect to the fixed basis b_i, the e_i are parametrized by the n×n matrix γ^{-1} consisting of n^2 numbers. However, any other basis obtained from e_i by a Lorentz rotation describes the same metric. Therefore we have to subtract from n^2 the number of dimensions of the Lorentz group, $\frac{1}{2}n(n-1)$, yielding again

$$n^2 - \tfrac{1}{2}n(n-1) = \tfrac{1}{2}n(n+1) \ . \tag{8.12}$$

Being nondegenerate, a metric g on a vector space V induces a canonical metric g^* on the dual vector space V^*: Let β^i be the basis of V^* dual to the basis b_i:

$$\beta^i(b_j) = \delta^i_j \ . \tag{8.13}$$

Define a metric on V^* by

$$g^*(\beta^i,\beta^j) = (g_{ij})^{-1} \ . \tag{8.14}$$

This metric is canonical, i.e. it does not depend on the choice of the basis b_i. It follows that the dual basis of an orthonormal basis e_i of V is itself orthonormal with respect to g^*, because η is its own inverse. *Attention*: In the following we denote an orthonormal basis of V^* by e^i, only the position of the index distinguishes basis from dual basis.

9. METRICS ON AN OPEN SUBSET OF $\mathbb{R}^n$

We defined a vector field on an open subset U of $\mathbb{R}^n$ as a differentiable family of vectors indexed by the points x of U. Likewise we now define a metric g on U to be a differentiable family g_x of vector space metrics. With repect to a frame $b_i(x)$ this family is described by the symmetric n×n matrix

$$g_{ij}(x) := g_x(b_i(x),b_j(x)) \ , \tag{9.1}$$

whose elements are real valued functions on U. For convenience we shall often suppress the x's in the following.

Since the orthonormalization procedure of Gram and Schmidt only involves addition, multiplication and division, that is differentiable operations, it also immediately guarantees the existence of orthonormal frames $e_i(x)$,

$$g_x(e_i(x), e_j(x)) = \eta_{ij} , \qquad (9.2)$$

with x-independent right-hand side.

A frame may now have two nice properties: being holonomic or being orthonormal. As often in life, we can have both only in trivial situations.

Theorem: An open subset U of $\mathbb{R}^n$ admits a holonomic and orthonormal frame if and only if it is flat.

We do not yet have a definition of flatness, but it is sufficient to take the naive sense of the word, for instance meaning that the angles of a triangle add up to 180°.

Let us return to our example of $\mathbb{R}^3$ minus a half plane and endow it with the Euclidean metric

$$g_{ij} = \begin{bmatrix} 1 & 0 & 0 \\ 0 & 1 & 0 \\ 0 & 0 & 1 \end{bmatrix} \qquad (9.3)$$

with respect to the Cartesian holonomic frame, which is therefore also orthonormal. On the other hand, the polar holonomic frame is not orthonormal:

$$g'^{ij} = (\gamma^{-1T}\, 1\, \gamma^{-1}) = \begin{bmatrix} 1 & 0 & 0 \\ 0 & r^2 \sin\theta & 0 \\ 0 & 0 & r^2 \end{bmatrix} , \qquad (9.4)$$

or, in the dual frame,

$$g^{ij} = \begin{bmatrix} 1 & 0 & 0 \\ 0 & 1 & 0 \\ 0 & 0 & 1 \end{bmatrix} \qquad (9.5)$$

and

$$g'^{ij} = \begin{bmatrix} 1 & 0 & 0 \\ 0 & r^{-2} \sin^{-2}\theta & 0 \\ 0 & 0 & r^{-2} \end{bmatrix} . \qquad (9.6)$$

To have a non-flat example consider a piece of the unit sphere, r=1. It is an open subset of $\mathbb{R}^2$ parametrized by φ and θ. Its metric is given by

$$g^{ij} = \begin{bmatrix} \frac{1}{\sin^2\theta} & 0 \\ 0 & 1 \end{bmatrix} \qquad (9.7)$$

with respect to the holonomic frame $d\varphi$, $d\theta$. An orthonormal frame is, for instance,

$$e^1 = \sin\theta \, d\varphi \ , \ e^2 = d\theta \ . \qquad (9.8)$$

It is not holonomic:

$$de^1 = d(\sin\theta \, d\varphi) = \cos\theta \, d\theta \wedge d\varphi \neq 0 \ . \qquad (9.9)$$

Of course, this does not exclude the existence of some other holonomic orthonormal frame, and only in the last chapter shall we be able to decide that this space is in fact not flat.

10. THE HODGE STAR

The Hodge star is a map turning a p-form into an (n-p)-form. We define it in terms of a holonomic frame:

$$* : \Lambda^p U \to \Lambda^{n-p} U$$

$$\phi \mapsto *\phi$$

$$*\phi := \frac{1}{(n-p)!} \sum_{\mu_{p+1}\dots\mu_n} \Big[\frac{1}{p!} \sum_{\mu_1\dots\mu_p} \epsilon_{\mu_1\dots\mu_n} \sqrt{|\det(g_{kl})|}$$

$$\times \sum_{\nu_1\dots\nu_p} \phi_{\nu_1\dots\nu_p} g^{\mu_1\nu_1} \dots g^{\mu_p\nu_p} \Big] dx^{\mu_{p+1}} \wedge \dots \wedge dx^{\mu_n} \ , \qquad (10.1)$$

where $\epsilon_{\mu_1\dots\mu_n}$ is the completely antisymmetric tensor with

$$\epsilon_{1\dots n} = 1 \ . \qquad (10.2)$$

Note that this definition requires the choice of an orientation in $\mathbb{R}^n$, but does not depend on the particular coordinate system used. Like the wedge product, the Hodge star is a purely algebraic operation. It is linear and its square is plus or minus the identity:

$$**\phi = (-1)^{p(n-1)+s} \phi . \tag{10.3}$$

Recall that s is the number of minus signs in the metric. Note that the Hodge star has a particularly simple expression in an orthonormal frame.

11. THE CODERIVATIVE

Just like the exterior derivative, the coderivative is a linear first order differential operator, which, however, lowers the degree of a differential form by one unit:

$$\delta : \Lambda^p U \to \Lambda^{p-1} U$$

$$\phi \mapsto \delta\phi$$

$$\delta\phi := (-1)^{np+n+1+s} * d * \phi . \tag{11.1}$$

From the exterior derivative it also inherits the nilpotency:

$$\delta^2 = 0 . \tag{11.2}$$

12. THE LAPLACE OPERATOR

The Laplace operator is the linear second order differential operator defined by:

$$\Delta : \Lambda^p U \to \Lambda^p U$$

$$\Delta := - (d\delta + \delta d) . \tag{12.1}$$

If the metric is indefinite, it is usually called the d'Alembert operator, and written as a box $\Box$.

13. SUMMARY OF THE MATHEMATICAL PART

Before starting the physical part, let us summarize:

We have recast a part of tensor analysis in a coordiante free language using differential forms. This serves two purposes:

- it has less indices, making some calculations more transparent.
- it can easily be generalized to more general spaces like manifolds.

The following dictionary may be useful:

v	v^{μ}
$\phi \in \Lambda^p U$	$\phi_{[\mu_1 \dots \mu_p]}$
$\phi(v_1, v_2, \dots, v_p)$	$\Sigma_{\mu_1 \dots \mu_p} \phi_{[\mu_1 \dots \mu_p]} v_1^{\mu_1} \dots v_p^{\mu_p}$
$\phi \wedge \psi$	$\phi_{[\mu_1 \dots \mu_p} \psi_{\mu_{p+1} \dots \mu_q]}$
$d\phi$	$\partial_{[\mu_1} \phi_{\mu_2 \dots \mu_{p+1}]}$
$\int_K \phi$	$\int_K \phi_{1 \dots p}\, dx^1 \dots dx^p$
g	$g_{(ij)}$
g^*	$g^{ij} = (g^{-1})_{ij}$
$*\phi$	$\Sigma_{\substack{\mu_1 \dots \mu_p \\ \nu_1 \dots \nu_p}} \sqrt{\lvert g \rvert}\, \phi_{\nu_1 \dots \nu_p} g^{\nu_1 \mu_1} \dots g^{\nu_p \mu_p} \epsilon_{\mu_1 \dots \mu_p, \mu_{p+1} \dots \mu_n}$
$-d*d* - *d*d$	Δ .

We first consider Minkowski space $U = \mathbb{R}^4$ equipped with the Minkowski metric of signature +---. We also adopt from now on Einstein's summation convention (summing over indices that appear twice) and put the speed of light equal to one: c=1. The sources, electric charge and current densities, are combined into a real valued 3-form:

$$j = \frac{1}{3!} \epsilon_{\mu\nu\lambda\rho} j^\mu dx^\nu \wedge dx^\lambda \wedge dx^\rho \in \Lambda^3(\mathbb{R}^4) . \qquad (14.1)$$

Integrating j over a 3-dimensional space-like volume yields the total charge inside that volume as a function of time. Charge conservation reads

$$dj = 0 . \qquad (14.2)$$

The electric and magnetic fields are written as a real valued 2-form

$$F = \tfrac{1}{2} F_{\mu\nu} dx^\mu \wedge dx^\nu , \qquad (14.3)$$

with

$$F_{\mu\nu} = \begin{bmatrix} 0 & -E_1 & -E_2 & -E_3 \\ E_1 & 0 & B_3 & -B_2 \\ E_2 & -B_3 & 0 & B_1 \\ E_3 & B_2 & -B_1 & 0 \end{bmatrix} . \qquad (14.4)$$

Maxwell's equations now read (in Heaviside units):

$$dF = 0 , \qquad (14.5)$$

$$\delta F = *j . \qquad (14.6)$$

Eq. (14.6) implies charge conservation. Therefore only conserved currents, dj = 0, may be coupled to the electromagnetic field. Our space-time being simply connected, Eq. (14.5) implies the existence of a potential, a real valued 1-form A such that

$$F = dA . \qquad (14.7)$$

Expressed in terms of the potential, Eq. (14.6) can be obtained from the action

$$S = \int (- \tfrac{1}{2} F \wedge *F - j \wedge A) \qquad (14.8)$$

upon variation of the potential. This means that we replace A by A+a in the action, expand it, and put the term linear in a equal to zero.

The advantages of this formulation are twofold:

- Lorentz invariance is immediate; SO(1,3), the group of linear transformations preserving the metric and the orientation of $\mathbb{R}^4$, also leaves the Hodge star, and consequently the Maxwell action (14.8), invariant.

- In Maxwell's equations, or in the action, the flat Minkowski metric may be replaced by any curved metric, indicating how electromagnetism is coupled to gravity.

We further note that the potential is not unique, indeed

$$A' = A + d\Lambda, \qquad \Lambda \in \Lambda^0(\mathbb{R}^4) \tag{14.9}$$

implies

$$F' = F. \tag{14.10}$$

Therefore, and because of charge conservation, these gauge transformations leave the action invariant. The underlying symmetry group is the set of functions from space-time into U(1) denoted by ${}^{\mathbb{R}^4}U(1)$. That the relevant group is really U(1) and not $\mathbb{R}$ can be seen when electromagnetism is ("minimally") coupled to a quantum system, for example to Schrödinger's equation. Then the U(1) becomes precisely the group of phase transformations.

15. GAUGE THEORIES AND THE YANG-MILLS ACTION

We now generalize the U(1) in Maxwell's theory to a bigger nonabelian group. Let G be a finite dimensional real Lie group, for example SU(2), SO(1,3) or GL_4. For simplicity we shall only consider matrix groups. We denote by $\mathcal{G}$ the Lie algebra of G. For the matter fields we also need a linear representation ρ of G on a real vector space W, for example in the case of SU(2) the fundamental representation with $W = \mathbb{C}^2$ considered as a 4-dimensional real vector space, or $W = \mathbb{R}^4$ for SO(1,3). We denote by $\tilde{\rho}$ the corresponding linear representation of the Lie algebra $\mathcal{G}$ on W. A gauge transformation γ is a map from spacetime into G

$$\begin{aligned} \gamma : U &\to G \\ x &\mapsto \gamma(x) \ . \end{aligned}$$

The set of all these maps forms the gauge group ${}^{U}G$, an infinite dimensional group, in general not a Lie group[1].

We introduce a connection A, a 1-form on U with values in the Lie algebra $\mathcal{G}$:

$$A \in \Lambda^1(U, \mathcal{G}) \ . \tag{15.1}$$

It is also called a potential or gauge field. For SU(2), for instance, A is a 2×2 matrix

$$A = \begin{bmatrix} i\alpha_3 & \alpha_2+i\alpha_1 \\ -\alpha_2+i\alpha_1 & -i\alpha_3 \end{bmatrix}, \tag{15.2}$$

where the coefficients α_a are real valued 1-forms. In general, if T_a, $a=1,2,\ldots,\dim G$, is a basis of $\mathfrak{g}$, the connection can be expanded as:

$$A = \sum_{a=1}^{\dim G} A^a\, T_a = \sum_a \sum_\mu A^a_\mu\, dx^\mu\, T_a\,, \tag{15.3}$$

with
$$A^a \in \Lambda^1(U,\mathbb{R}) \tag{15.4}$$

and real valued functions A^μ_a. Under a gauge transformation $\gamma \in {}^U G$ the connection is required to transform inhomogeneously:

$$A' = \gamma\, A\, \gamma^{-1} + \gamma\, d\gamma^{-1}\,. \tag{15.5}$$

We define the field strength, or curvature F, to be the Lie algebra valued 2-form

$$F := dA + \tfrac{1}{2}\,[A,A] \in \Lambda^2(U,\mathfrak{g})\,. \tag{15.6}$$

Its gauge transformation is easily seen to be homogeneous:

$$F' = \gamma\, F\, \gamma^{-1}\,. \tag{15.7}$$

Note that only for abelian (internal) gauge groups is the field strength gauge invariant, and may therefore be a physical observable.

For simplicity we only consider spin 0 matter fields, 0-forms with values in the representation space W:

$$\phi \in \Lambda^0(U,W)\,. \tag{15.8}$$

For example, in the case of the fundamental representation of SU(2):

$$\phi = \begin{bmatrix} \phi_1+i\phi_2 \\ \phi_3+i\phi_4 \end{bmatrix}, \qquad \phi_i \in \Lambda^0(U,\mathbb{R})\,. \tag{15.9}$$

The matter fields transform under gauge transformations according to their representation ρ:

$$\phi' = \rho(\gamma)\,\phi\,. \tag{15.10}$$

The covariant (exterior) derivative is defined by

$$D\phi := d\phi + \tilde{\rho}(A) \wedge \phi \in \Lambda^1(U,W)\,. \tag{15.11}$$

Just like the field strength, $D\phi$ is easily seen to transform homogeneously:

$$(D\phi)' = \rho(\gamma)\, D\phi\,. \tag{15.12}$$

The nonabelian generalization of the first Maxwell equation (14.5), now called the Bianchi identity, is an immediate consequence of the definition (15.6):

$$DF = dF + [A,F] = 0 \ . \qquad (15.13)$$

Furthermore,

$$D^2\phi = \tilde{\rho}(F) \wedge \phi \ . \qquad (15.14)$$

Mimicing the Maxwell action, we define the Yang-Mills action

$$S = \int \left[\frac{1}{g^2} \operatorname{tr}(F \wedge F) + D\phi^+ \wedge *D\phi - m^2\phi^+ \wedge *\phi \right] . \qquad (15.15)$$

In four dimensions the Yang-Mills coupling constant g is dimensionless.

The Yang-Mills action is invariant under the huge infinite dimensional gauge group. Consequently, the field equations obtained from this action by varying the connection A will be gauge invariant. These field equations are the Yang-Mills equations. There are quite a number of exact solutions to these non-linear second order differential equations. However, their physical interpretation remains unclear until today. The success of Yang-Mills theories lies so far exclusively in the quantum sector.

Variation of the Yang-Mills action with respect to the matter fields yields the gauge covariant generalization of the Klein-Gordon equation.

16. THE EQUIVALENCE PRINCIPLE

Up to now the space-time metric was fixed *a priori*. It seems natural at this point to try to promote the metric to a dynamical field (describing gravitational interaction). To this end we look for differential equations determining the metric. By definition the metric is a differentiable family of bilinear symmetric forms, and we do not know what differential equations for such objects are (the existence of the exterior derivative for differential forms is intimately linked to their antisymmetry). We have already seen that any metric can be described by orthonormal frames of 1-forms. For these we know differential operators. The equivalence principle requires that the metric and only the metric generates gravitational interaction, in particular that the frame chosen to describe the metric is irrelevant. Our task, therefore, is to find differential equations for the orthonormal frames e^i which are covariant under gauge transformations Λ :

$$e'^i = \Lambda^i{}_j e^j \ , \qquad \Lambda \in {}^{U}SO(1,3) \ . \qquad (16.1)$$

We restrict ourselves to orientation preserving Lorentz transformations because we want to use the Hodge star later on. It is sometimes convenient to consider the orthonormal frame e^i as a 1-form e with values in the fundamental representation of SO(1,3). To be more precise, we must add the restriction that the e^i be linearly independent, which is compatible with the gauge transformation

$$e' = \Lambda e . \tag{16.2}$$

To get gauge covariant field equations for e we use the same trick as before: We introduce a connection, write down an invariant action and obtain the desired field equations by variation. In Yang-Mills theories the connection actually represents new physical fields, like the photon or the weak bosons $W^{\pm}$,Z. Here we just signed the equivalence principle prohibiting the introduction of new fields. A natural solution to this dilemma will show up automatically, and for the moment we allow for a new field, the connection ω, a 1-form with values in the Lie algebra of SO(1,3)

$$\omega \in \Lambda^1(U,so(1,3)) , \tag{16.3}$$

also called spin connection. As a connection, it is supposed to transform under gauge transformations according to

$$\omega' = \Lambda \, \omega \, \Lambda^{-1} + \Lambda \, d\Lambda^{-1} . \tag{16.4}$$

As before, we define the curvature

$$R := d\omega + \tfrac{1}{2} [\omega,\omega] \quad \in \quad \Lambda^2(U,so(1,3)) . \tag{16.5}$$

This definition is known as Cartan's second structure equation. Again we have immediately the homogeneous transformation property of the curvature:

$$R' = \Lambda R \Lambda^{-1} . \tag{16.6}$$

We define torsion by Cartan's first structure equation

$$T := de + \omega \wedge e = De \in \Lambda^1(U,\mathbb{R}^4) . \tag{16.7}$$

As a covariant derivative, also the torsion transforms homogeneously under gauge transformations:

$$T' = \Lambda T . \tag{16.8}$$

As before (cf. (15.13),(15.14)), we have the Bianchi identities:

$$DR = dR + [\omega,R] = 0 . \tag{16.9}$$

$$DT = R \wedge e . \tag{16.10}$$

For a Yang-Mills theory without matter the cheapest gauge invariant action is quadratic in the curvature:

$$S = \frac{1}{g^2} \int F^a{}_b \wedge *F^b{}_a \,. \tag{17.1}$$

For the moment the pure gravitational field is coded into two fields; e and ω. Consequently, we have an invariant action linear in the curvature,

$$S = \frac{-1}{32\pi G} \int R^a{}_b \wedge *(e^b \wedge e_a) \,, \tag{17.2}$$

where from now on indices are raised and lowered with η^{ab} and η_{ab} and e always denotes orthonormal frames of 1-forms. Eq. (17.2) is the Einstein-Hilbert action. Its gravitational coupling constant has units $(\text{meter})^2$ in four dimensions. Using the definition of the Hodge star in four dimensions the Einstein-Hilbert action can also be written as

$$S = \frac{-1}{32\pi G} \int R^{ab} \wedge e^c \wedge e^d \, \epsilon_{abcd} \,. \tag{17.3}$$

We introduce matter by adding a functional $\int \mathcal{L}_M$ depending on the matter fields, e and ω,

$$S = \frac{-1}{32\pi G} \int R^{ab} \wedge e^c \wedge e^d \, \epsilon_{abcd} + \int \mathcal{L}_M[e,\omega] \,. \tag{17.4}$$

For example, the matter could be a Yang-Mills action (15.15) with A and φ now considered as matter fields. This particular matter action depends only on e (through the Hodge star) and not on ω.

Let us derive the field equations following from (17.4).

<u>Variation of e</u>

We call τ the variation of the matter Lagrangian with respect to e:

$$\mathcal{L}_M[e+f] - \mathcal{L}_M[e] =: -f^c \wedge \tau_c + O(f^2) \,, \tag{17.5}$$

where τ is a 3-form with values in $\mathbb{R}^4$,

$$\tau \in \Lambda^3(U,\mathbb{R}^4) \,, \tag{17.6}$$

the "energy momentum tensor". Integrating τ over a 3-dimensional volume yields the energy momentum contained in that volume. E.g. for pure electromagnetic radiation,

$$\mathcal{L}_M = -\tfrac{1}{2} F \wedge *F \tag{17.7}$$

and we obtain, after a lengthy calculation,

$$\tau_{00} = \tfrac{1}{2} \left(\vec{E}^2 + \vec{B}^2 \right) \,, \tag{17.8}$$

with

$$*\tau_c =: \tau_{ca} e^a \,. \tag{17.9}$$

Variation of the total action (17.4) with respect to e immediately gives the Einstein equations:

$$R^{ab} \wedge e^d \epsilon_{abcd} = -16\pi G \tau_c \, . \qquad (17.10)$$

For given energy momentum τ, they are non-linear first order differential equations for the connection. They are also linear equations for the curvature, "energy is the source of curvature". Despite the algebraic nature of the equations curvature propagates in four dimensions: Vanishing τ does not imply vanishing curvature, as illustrated, for example, by Schwarzschild's solution. This comes from the fact that the curvature has $6\times 6 = 36$ independent coefficients $R^{ab}{}_{\mu\nu}$ (antisymmetric in μ and ν because R is a 2-form, antisymmetric in a and b because R takes values in the Lorentz algebra) while Einstein's equation, being an equation for 3-forms with values in $\mathbb{R}^4$, contains only $4\times 4 = 16$ linear equations. In two- and three-dimensional space-times the counting is different and curvature does not propagate.

Variation of ω

We define the spin density

$$S \in \Lambda^3(U, so(1,3)) \qquad (17.11)$$

by

$$\mathcal{L}_M[\omega+\chi] - \mathcal{L}_M[\omega] =: -\tfrac{1}{2}\, \chi^{ab} \wedge S_{ab} + O(\chi^2) \, . \qquad (17.12)$$

Of course, the spin density is zero for the Yang-Mills action (15.15). It is non-vanishing, for instance, for the Dirac action describing spin $\frac{1}{2}$ fields, which motivates the name spin density. Varying ω in the total action (17.4) yields, after an integration by parts, the equation

$$T^c \wedge e^d \epsilon_{abcd} = -8\pi G\, S_{ab} \, , \qquad (17.13)$$

"spin is the source of torsion". If we now count the number of linear equations and unknowns, we find them to match in any dimension. Torsion does not propagate: Vanishing spin density implies vanishing torsion.

We now come to the promised elimination of the spin connection as an independent field. There are two possible routes.

Einstein's point of view

Einstein puts torsion to zero right from the beginning. By virtue of equation (16.7),

$$0 = T = de + \omega \wedge e \qquad (18.1)$$

is a covariant constraint and therefore it does not spoil the covariance of Einstein's equation. Let us consider the constraint (18.1) as a system of linear equations with the components of the spin connection $\omega^{ab}{}_{\mu}$ as unknowns. Since ω is so(1,3)-valued, it is antisymmetric in the indices a and b, and there are 6×4 unknowns. On the other hand, (18.1) is an equation for $\mathbb{R}^4$-valued 2-forms and has 4×6 components $T^a{}_{\mu\nu}$. Consequently, there exists (for any signature and dimension) a unique solution expressing the spin connection as a function of the frame and its first derivatives. This solution is called a Riemannian connection. Its explicit form is most conveniently written down by expanding ω with respect to the orthonormal frame e:

$$\omega^a{}_b = \omega^a{}_{bc} e^c \; . \qquad (18.2)$$

Then the Riemannian connection is given by

$$\omega^a{}_{bc} = \tfrac{1}{2} \left(C^a{}_{bc} - C_b{}^a{}_c - C_c{}^a{}_b \right) , \qquad (18.3)$$

where the functions C are defined by

$$de^a =: \tfrac{1}{2} C^a{}_{bc} \, e^b \wedge e^c \; . \qquad (18.4)$$

Substituting the Riemannian connection ω(e,∂e) into Einstein's equations they become non-linear second order differential equations for the orthonormal frame. Alternatively, they can be obtained by substituting first the Riemannian connection into the Einstein-Hilbert action and then varying with respect to the frame ("second order formalism").

Cartan's point of view

Cartan keeps ω as an independent field which eliminates itself at the end through its own (algebraic) field equation (17.13): ω = ω(e,∂e,s). Therefore in this so-called Einstein-Cartan theory the equivalence principle is only valid outside matter with spin. Only there is it verified experimentally. Furthermore, the observed spin density in the universe is small and torsion couples to it via the universal coupling constant G, implying that although different in principle the Einstein and Einstein-Cartan theories are presently indistinguishable experimentally.

It can be shown[2] that the Einstein-Hilbert action is the unique action that leads to vanishing torsion in the vacuum as a field equation, unique, of course, up to terms containing no spin connection, e.g. the cosmological term

$$\frac{\Lambda}{4!} \int e^a \wedge e^b \wedge e^c \wedge e^d \epsilon_{abcd} \cdot \tag{18.5}$$

The Einstein-Cartan theory with a spin 3/2 Rarita-Schwinger field has in addition to its SO(1,3) gauge invariance one more remarkable invariance, supersymmetry[3,4] and has become known as supergravity. On the other hand, Einstein's general relativity with vanishing torsion does not seem to admit supersymmetry.

As promised, we now show that a piece of the 2-dimensional unit sphere (chapter 9) cannot have a holonomic and orthonormal frame.

Theorem: An open subset U of $\mathbb{R}^n$ with a metric g admits a holonomic and orthonormal frame if and only if its Riemannian connection has everywhere vanishing curvature.

We use equation (18.3) to calculate the Riemannian connection from

$$e^1 = \sin\theta \, d\varphi \, , \quad e^2 = d\theta \tag{18.6}$$

and

$$de^1 = -\frac{\cos\theta}{\sin\theta} e^1 \wedge e^2 \tag{18.7}$$

$$d^2 e = 0 \, . \tag{18.8}$$

Therefore

$$C^1{}_{12} = -C^1{}_{21} = -\frac{\cos\theta}{\sin\theta} \, , \tag{18.9}$$

all other C's being zero. Consequently, the Riemannian connection is

$$\omega^1{}_2 = -\frac{\cos\theta}{\sin\theta} e^1 = \cos\theta \, d\varphi \, , \tag{18.10}$$

and its curvature

$$R^1{}_2 = e^1 \wedge e^2 \tag{18.11}$$

is different from zero.

To conclude, following Cartan we have presented general relativity using orthonormal frames. This may be somewhat unfamiliar because Einstein formulated his theory with the help of holonomic frames. Of course, both approaches have advantages and disadvantages. Three major shortcomings of holonomic frames are:

- Their invariance group is GL_4, which does not admit spinor representations[5], therefore excluding fields with half integer spin.
- They break the gauge invariance of general relativity, ignoring today's belief that all fundamental interactions are described by gauge theories.
- The treatment of gravitational anomalies is quite awkward in holonomic frames.

For further details and literature we refer to Göckeler & Schücker[6].

REFERENCES

1. J. Milnor, in: "Relativity, Groups and Topology II", B. De Witt and R. Stora, eds., North Holland, Amsterdam (1984).
2. R. G. Yates, Comm. Math. Phys. 76:255. (1980).
3. S. Deser and B. Zumino, Phys. Lett. 62B:335.(1976).
4. D. Freedman, P. van Nieuwenhuizen, and S. Ferrara, Phys. Rev. D13:3214 (1976).
5. E. Cartan, "Lecons sur la théorie des spineurs", Hermann, Paris (1938).
6. M. Göckeler and T. Schücker, "Differential Geometry, Gauge Theories and Gravity", Cambridge University Press (1987).

QUANTIZATION OF CONSTRAINED SYSTEMS

N. K. Falck*

II. Institut für Theoretische Physik
der Universität Hamburg
Notkestr. 85, 2000 Hamburg 52, Fed. Rep. Germany

ABSTRACT

This lecture presents Dirac's quantization procedure for constrained systems, modified such that gauge degrees of freedom are eliminated by gauge fixing. The application of the Dirac bracket quantization procedure to Maxwell's electromagnetism serves as an example. In a second lecture the method is applied in a discussion of the gauge invariance of the chiral Schwinger model.

* Supported by the Bundesministerium für Forschung und Technologie, 05 4HH 928/3, Bonn FRG

1. INTRODUCTION

In the present lecture I'm going to present a pretty old subject, which, nevertheless, up to now did not receive the attention it deserves; the quantization procedure for constrained systems[1,2]. Especially in elementary particle physics constrained systems are extremely important since, according to our present understanding, all fundamental interactions are mediated by gauge particles (photons, W and Z bosons, gluons and gravitons). The corresponding gauge theories for electroweak, strong and gravitational interactions all belong to the class of constrained systems, a feature which is associated with the presence of gauge invariance.

2. GENERAL ANALYSIS

Nowadays it is usual to define a theory in terms of its classical action

$$S = \int dt\ L(q^i,\dot{q}^i)\ , \tag{1}$$

where q^i, $i = 1,\ldots,d$ are coordinates of the configuration space, $\dot{q}^i$ the corresponding velocities and L is the Lagrange function. However, there is no reliable quantization procedure which does not make use of the Hamilton formalism (even in Feynman's path integral quantization, difficult questions like the behaviour under point transformations can be answered only in the Hamilton formalism). Therefore we are forced to perform the Legendre transformation in order to find the Hamiltonian. This is a transformation of the type

$$(q^i,\dot{q}^i) \to (q^i,p_i)\ , \tag{2}$$

where the canonical momenta p_i are defined by:

$$p_i = \frac{\delta L}{\delta \dot{q}^i}\ . \tag{3}$$

The derivative in Eq. (3) is a functional derivative if the theory under consideration is a field theory. In this case the index i is continuous and represents also the space point. The Jacobi matrix of this transformation reads

$$J_{ij} = \frac{\delta p_i}{\delta \dot{q}^j} = \frac{\delta^2 L}{\delta \dot{q}^i \, \delta \dot{q}^j} \, . \tag{4}$$

As long as $\det J \neq 0$ the Legendre transformation does not present any difficulties. We are just interested, however, in the other case, where $\det J = 0$. Such theories are called constrained. $\det J = 0$ implies that some of the momenta are not independent; they are determined by equations of the form

$$\Phi_m(q^i, p_i) \approx 0, \; m = 1,\ldots,M \, , \tag{5}$$

where the symbol "$\approx$" means "weakly equal", this will be explained later. Eqs. (5) are called "primary constraints", constraints because they don't involve the time evolution and primary since they follow directly from the canonical procedure. The Hamilton function is usually constructed from the requirement that it does not depend on the velocities. In the present case this is not unique, since the primary constraints do not depend on $\dot{q}^i$ either. They can be added to the canonical Hamiltonian (we use Einstein's summation convention)

$$H = p_i \dot{q}^i - L \, , \tag{6}$$

with arbitrary coefficients (Lagrange multipliers) u_m, to give the "total" Hamiltonian

$$H_T = H + u_m \Phi_m \, . \tag{7}$$

The Hamiltonian equations of motion read

$$\dot{q}^i = \frac{\delta H}{\delta p_i} + u_m \frac{\delta \Phi_m}{\delta p_i} \, , \tag{8}$$

$$\dot{p}_i = - \frac{\delta H}{\delta q^i} - u_m \frac{\delta \Phi_m}{\delta q^i} \, . \tag{9}$$

The arbitrariness of the dynamics reflected by the Lagrange multipliers can be elucidated by a simple example: consider a particle of unit mass moving on a circle in a plane. The corresponding Lagrangian reads:

$$L = \tfrac{1}{2} \dot{\vec{q}}^2 - V(\vec{q}) - \lambda(\vec{q}^2 - r^2) \, . \tag{10}$$

The total configuration space contains three coordinates: q^1, q^2 and λ. It is important to consider λ as a degree of freedom, since it is the Euler-Lagrange equation with respect to λ which forces the particle onto the circle. However, since L does not contain $\dot{\lambda}$, the velocity of λ remains undetermined. The canonical momenta are

$$p_1 = \dot{q}^1 ; \quad p_2 = \dot{q}^2 ; \quad p_\lambda \approx 0 \tag{11}$$

and the canonical Hamiltonian reads

$$H = {}^1/_2\, \vec{p}^2 + V(\vec{q}) + \lambda(\vec{q}^2 - r^2) . \tag{12}$$

This implies $\dot{\lambda} = \partial H/\partial p_\lambda = 0$, which does not fit to the original Lagrangian system. This mismatch is cured by a Lagrange multiplier u, associated to the primary constraint $\phi = p_\lambda \approx 0$, in passing over to the total Hamiltonian:

$$H_T = H + u \cdot p_\lambda . \tag{13}$$

Hence we find $\dot{\lambda} = u$, which is arbitrary, as it should be.

The equations of motion (8,9) may be rewritten using the Poisson bracket

$$\{f,g\} = \frac{\delta f}{\delta q^i}\frac{\delta g}{\delta p_i} - \frac{\delta f}{\delta p_i}\frac{\delta g}{\delta q^i} \tag{14}$$

according to

$$\dot{f} \approx \{f,H_T\} + \frac{\partial f}{\partial t} . \tag{15}$$

For Eq. (15) to reproduce Eqs. (8,9), it is important that the constraint equations are not used before the Poisson brackets are evaluated; this is indicated by the notion of weak equality. The term $\{f,u_m\}\phi_m$ in Eq. (15), though not defined, is harmless, since the ϕ_m are going to vanish anyway.

This is not the whole story: the constraints have to be fulfilled not only at some fixed time, but for all times, which means that consistency requires

$$\dot{\phi}_m \approx \{\phi_m, H_T\} \overset{!}{\approx} 0 . \tag{16}$$

This is not only necessary but also desired. In the example we are here considering $\dot{\phi} = \dot{p}_\lambda = - \delta H/\delta\lambda \approx 0$ means that the particle is forced to move on the circle, which is very good, since this was the reason to introduce λ in the first place. There are four kinds of consequences which the imposition of the consistency conditions of Eq. (16) can have:

(i) They lead to an inconsistency of the form 0=1. Then we can forget about the theory, hence we shall assume that this does not happen.

(ii) Some of the conditions may be fulfilled automatically, this does not give any new information.

(iii) The conditions give new, so-called secondary, constraints

$$\chi(p,q) \approx 0 \ . \tag{17}$$

Then the procedure has to be repeated: require χ to be conserved as well. This can lead to a number, say N, of secondary constraints.

(iv) Conditions containing the Lagrange multipliers can be solved, this fixes some of the Lagrange multipliers.

After having finished this procedure, we end up with $M + N$ constraints

$$\chi_k(q,p) \approx 0, \quad k = 1,\dots,M+N \tag{18}$$

and the "extended" Hamiltonian

$$H_E = H' + \sum_{a=1}^{A} v_a \chi_a(q,p) \ , \tag{19}$$

where H' is defined by the solutions $\mho$ of the consistency requirements of case (iv):

$$H' = H + \mho_m \chi_m(q,p) \tag{20}$$

and v_a, $a=1,\dots,A$ are the leftover undetermined Lagrange multipliers. The sum in Eq. (19) has to include, not only primary, but also secondary first class constraints (see below). Now we can define the following classification: a quantity is called first class if it has vanishing Poisson brackets with all constraints, otherwise it is called second class. The extended Hamiltonian is first class since the constraints are conserved by construction. The constraints denoted by χ_a in Eq. (19) are first class too, because otherwise the conditions $\dot{\chi}_k \approx 0$ would determine the v_a, which contradicts our assumption that the v_a are not fixed. All constraints contained in H' in Eq. (20) are second class; this is necessary in order to fix $\mho_m$ by virtue of the consistency conditions $\dot{\chi}_k \approx \{\chi_k, H_E\} \approx 0$.

Before proceeding to quantization, the meaning of the first class constraints should be clarified. Consider some dynamical quantity $g(t)$ with initial value $g(0) = g_0$. After an infinitesimal time δt has passed, g acquires the value

$$g(\delta t) = g_0 + \dot{g}\cdot\delta t \approx g_0 + (\{g,H'\} + v_a\{g,\chi_a\})\delta t \ , \qquad (21)$$

which depends on the arbitrary parameters v_a. If we had chosen another set of Lagrange multipliers, say v_a', we would have found

$$g'(\delta t) \approx g_0 + (\{g,H'\} + v_a'\{g,\chi_a\})\delta t \ . \qquad (22)$$

Thus the variation with respect to the Lagrange multipliers is

$$\Delta g(\delta t) = \epsilon_a \{g,\chi_a\} \ , \qquad (23)$$

where $\epsilon_a = \delta t\cdot(v_a - v_a')$ are arbitrary infinitesimal local parameters. This means that first class constraints are infinitesimal generators of local transformations which do not alter the physics, since g is as good as g' (physics does not care about a theorist's choice of Lagrange multipliers). Transformations of this kind are called gauge transformations, especially in field theories.

Dirac proposed two different quantization procedures, corresponding to first class and second class constraints[1]. In the meantime, however, it has been observed[3] that only one of these procedures is necessary, since first class constraints can be converted to second class ones by imposing gauge conditions of the form

$$\chi_a \approx 0 \ , \quad a = A+1,\ldots 2A \qquad (24)$$

i.e. one gauge condition for each first class constraint. Then we have $M+N+A$ second class constraints χ_r, provided that the determinant of the matrix $\{\chi_a,\chi_b\}$ does not vanish and no first class constraint survives gauge fixing.

Second class constraints introduce a problem upon quantization; it is not possible to apply the usual procedure to transcribe Poisson brackets into commutators. For $\{\chi_r,\chi_s\} \neq 0$ would translate into $[\chi_r,\chi_s] = \chi_r\chi_s - \chi_s\chi_r \neq 0$, which makes it impossible to impose the constraints at the quantum level: the constraints can be fulfilled neither as operator equations nor as conditions on the states. This difficulty can be overcome by introducing Dirac brackets, defined by:

$$\{f,g\}_D = \{f,g\} - \{f,\chi_r\}C_{rs}\{\chi_s,g\} , \quad (25)$$

where C is the inverse matrix of $\{\chi,\chi\}$:

$$C_{rs}\{\chi_s,\chi_t\} = \delta_{rt} . \quad (26)$$

C exists if all constraints are second class[1], which eventually has to be enforced by gauge fixing, as mentioned above. The Dirac bracket is antisymmetric and fulfills the Jacobi identity

$$\{f,\{g,h\}_D\}_D + \{g,\{h,f\}_D\}_D + \{h,\{f,g\}_D\}_D = 0 . \quad (27)$$

All Dirac brackets involving a constraint vanish:

$$\{f,\chi_t\}_D = \{f,\chi_t\} - \{f,\chi_r\}C_{rs}\{\chi_s,\chi_t\} = \{f,\chi_t\} - \{f,\chi_r\}\delta_{rt} = 0 \quad (28)$$

and all Dirac brackets involving first class quantities are identical to the corresponding Poisson brackets, since the additional terms in Eq. (25) vanish. Therefore the Hamilton equations of motion (Eq.(15)) can be formulated in terms of Dirac brackets as well (f is assumed not to depend on time explicitly) :

$$\dot{f} = \{f,H_E\} = \{f,H_E\}_D = \{f,H\}_D . \quad (29)$$

All this means: once the fundamental Dirac brackets have been calculated it is possible to consider the constraint equations as strong equations, provided that in the sequel only Dirac brackets are used. Since now all constraints are strong there is no problem in the quantization procedure, the commutators are abstracted from the corresponding Dirac brackets and the classical constraint equations turn into strong operator equations:

$$\{f,g\}_D = h \quad \Rightarrow \quad [\hat{f},\hat{g}] = i\hbar\hat{h} \quad (30)$$

$$\chi_r(q,p) = 0 \quad \Rightarrow \quad \chi_r(\hat{q},\hat{p}) = 0 \quad (31)$$

$$\dot{f} = \{f,H\}_D \quad \Rightarrow \quad \dot{f} = \frac{1}{i\hbar}[\hat{f},\hat{H}]. \quad (32)$$

One remark should be made at this point; the statement that quantization is straightforward is too simple-minded in some cases. If any of the quantities in Eqs. (30)-(32) contain products of noncommutative operators, there is an operator ordering problem which is still unsolved for the general case[4].

This finishes the general analysis. For most of the cases of physical interest it provides a consistent framework for the canonical quantization of classical systems with constraints.

3. AN EXAMPLE: MAXWELL'S ELECTRODYNAMICS

As an illustration I wish to present the quantization procedure of the free electromagnetic field in the radiation gauge. In the following the notation of Bjorken and Drell[5] is used. The field strengths are

$$F_{\mu\nu} = \partial_\mu A_\nu - \partial_\nu A_\mu \tag{33}$$

and the Lagrange function is given by

$$L = \int \mathcal{L}(x)\, d^3x \; ; \quad \mathcal{L} = -\frac{1}{4} F_{\mu\nu}F^{\mu\nu} \; . \tag{34}$$

The canonical momenta can be calculated to be

$$\pi_0 = \frac{\delta L}{\delta \dot{A}^0} = 0 \; , \tag{35}$$

$$\pi_i = \frac{\delta L}{\delta \dot{A}^i} = F_{i0} = \partial_i A_0 - \partial_0 A_i \; . \tag{36}$$

Eq. (35) clearly defines a primary constraint:

$$\chi_1 = \pi_0 \approx 0 \; . \tag{37}$$

The canonical and total Hamiltonians read

$$H_C = \int \left(\tfrac{1}{2}\pi_i\pi_i + \tfrac{1}{4} F_{ij}F^{ij} - A_0\partial_i\pi^i \right) d^3x \; , \tag{38}$$

$$H_T = H_C + \int u_1\pi_0 \, d^3x \; . \tag{39}$$

Now we have to ensure consistency by requiring the constraint(s) to be stable in time. The first step is

$$\dot{\chi}_1 \approx \{\pi_0, H_T\} = -\frac{\delta H}{\delta A^0} = \partial_i\pi^i \overset{!}{\approx} 0 \; , \tag{40}$$

hence we have to impose the secondary constraint

$$\chi_2 = \partial_i \pi^i \approx 0 \; , \tag{41}$$

which is just Gauss' law. The next step is to calculate $\dot{\chi}_2$:

$$\dot{\chi}_2 \approx \partial_i\{\pi^i, H\} = \tfrac{1}{2}\, \partial_i(\partial_j F^{ji} - \partial_j F^{ij}) = 0 \; , \tag{42}$$

which vanishes automatically. Hence the chain of consistency requirements terminates and we are left with the two constraints χ_1 and χ_2, which are first class, since

$$\{\chi_1,\chi_2\} = 0 \; . \tag{43}$$

Finally, we have to add $\int u_2\chi_2 d^3x$ to the total Hamiltonian to get the extended Hamiltonian:

$$H_E = H_T + \int u_2\chi_2 \, d^3x \; . \tag{44}$$

Up to now the dynamics is not fixed, due to the Lagrange multipliers u_1 and u_2, hence we have to impose two gauge conditions. We choose the radiation gauge:

$$\chi_3 = \partial_i A^i \approx 0 \; , \tag{45}$$

$$\chi_4 = A^0 \approx 0 \; . \tag{46}$$

Now consistency has to be checked:

$$\dot{\chi}_3 \approx \{\partial_i A^i, H_E\} = \partial_i \frac{\delta H_E}{\delta \pi_i} = \partial_i \pi_i + \partial^i \partial_i A_0 - \partial^i \partial_i u_2 \approx \Delta(u_2 - A_0) \approx \Delta u_2 \overset{!}{\approx} 0 \tag{47}$$

$$\dot{\chi}_4 \approx \{A^0, H_E\} = \frac{\delta H_E}{\delta \pi_0} = u_1 \overset{!}{\approx} 0 \; . \tag{48}$$

These consistency requirements can be used to solve for the Lagrange multipliers u_1 and u_2 (via boundary conditions). Hence we may conclude: (i) the gauge conditions are consistent with the dynamics, (ii) the gauge is completely fixed by χ_3 and χ_4, since there is no arbitrariness left over in Eqs. (47) and (48). Therefore the total system can be formulated as

$$\begin{aligned} H &= \int \left(\tfrac{1}{2} \pi_i \pi_i + \tfrac{1}{4} F_{ij} F^{ij} \right) d^3x \; , \\ \chi_1 &= \pi_0 \quad ; \quad \chi_2 = \partial^i \pi_i \; , \\ \chi_3 &= \partial_i A^i \; ; \quad \chi_4 = A^0 \; . \end{aligned} \tag{49}$$

Note that all constraints have been dropped in the Hamiltonian, this is only allowed since we intend to use Dirac brackets to determine the dynamics. To this aim we have to invert the Poisson bracket matrix

$$\{\chi(x),\chi(y)\} = \begin{bmatrix} 0 & 0 & 0 & -1 \\ 0 & 0 & -\Delta(x) & 0 \\ 0 & \Delta(x) & 0 & 0 \\ 1 & 0 & 0 & 0 \end{bmatrix} \delta^3(x-y) \; . \tag{50}$$

This is possible because the gauge is admissible and complete. The result is

$$C(x,y) = \begin{bmatrix} 0 & 0 & 0 & 1 \\ 0 & 0 & 1/\Delta(x) & 0 \\ 0 & -1/\Delta(x) & 0 & 0 \\ -1 & 0 & 0 & 0 \end{bmatrix} \delta^3(x-y) , \tag{51}$$

such that

$$\int C_{rs}(x,y) \{\chi_s(y),\chi_t(z)\} d^3y = \delta_{rt}\delta^3(x-z) . \tag{52}$$

This can be used to calculate the fundamental Dirac bracket:

$$\{A^i(x),\pi_j(y)\} = \delta^i{}_j \; \delta^3(x-y)$$

$$- \int \{A^i(x),\chi_2(z_1)\} \; C_{23}(z_1,z_2) \; \{\chi_3(z_2),\pi_j(y)\} \; d^3z_1 \; d^3z_2$$

$$= (\delta^i{}_j + \frac{\partial^i\partial_j}{\Delta} (x)) \; \delta^3(x-y) , \tag{53}$$

all the other terms vanish. This is precisely the "transverse δ-function" which had to be introduced by hand in Bjorken and Drell:

$$\partial_i(x) \; (\delta^i{}_j + \frac{\partial^i\partial_j}{\Delta}(x)) \; \delta^3(x-y) = \partial^j(y) \; (\delta^i{}_j + \frac{\partial^i\partial_j}{\Delta}(x)) \; \delta^3(x-y) = 0 . \tag{54}$$

This automatically ensures that the Dirac bracket of A with π is consistent with transversality of the fields. The Fourier transform of Eq. (53) reads

$$(\delta^i{}_j + \frac{\partial^i\partial_j}{\Delta}) \; \delta^3(x-y) = \frac{1}{(2\pi)^3} \int d^3k \; (\delta^i{}_j + \frac{k^i k_j}{k^2}) \; e^{i\vec{k}(\vec{x}-\vec{y})} , \tag{55}$$

so that the corresponding quantum system is the usual one described in Bjorken and Drell:

$$H = \int (\tfrac{1}{2} \pi_i\pi_i + \tfrac{1}{4} F_{ij}F^{ij}) \; d^3x ,$$

$$\partial_i\pi_i = 0 ; \quad \partial_i A^i = 0 ,$$

$$[A_i(x),\pi_j(y)] = \frac{i\hbar}{(2\pi)^3} \int d^3k \; (-\delta_{ij} + \frac{k_i k_j}{k^2}) \; e^{i\vec{k}(\vec{x}-\vec{y})} . \tag{56}$$

REFERENCES

1. P. A. M. Dirac, Can. J. Math. 2: 125 (1950);

   "Lectures on Quantum Mechanics," Yeshiva Univ. Press, New York (1964).
2. A. J. Hanson, T. Regge and C. Teitelboim, Constrained Hamiltonian Systems, Accad. Nat. dei Lincei, Rome (1976).
3. E. S. Fradkin and G. A. Vilkoviski, CERN-TH 2332 (1977) (unpublished).
4. N. K. Falck and A. C. Hirshfeld, Eur. J. Phys. 4:5 (1983).
5. J. D. Bjorken and S. D. Drell, "Relativistic Quantum Fields," McGraw-Hill, New York (1965).

G-SPACES AND KALUZA-KLEIN THEORY

N. A. Papadopoulos

Institut für Physik der Johannes Gutenberg Universität
Postfach 3980, Staudinger Weg 7, 6500 Mainz, Fed. Rep. Germany

ABSTRACT

G-spaces are present whenever symmetries are relevant in physics. After a short introduction to this subject, spontaneous symmetry breaking in elementary particle physics is considered from this point of view. Kaluza-Klein theory is discussed in a purely geometrical formulation. Some results in connection with the geometrical compactification scheme are presented.

1. INTRODUCTION

It is not unusual, and probably natural, to work in physics for a long time with various objects, performing different calculations, and to realize only later that these have a very general mathematical or geometrical structure and play, even in mathematics, a central role.

It seems to me that this is the case with G-spaces[1]. We meet them in physics every day but we do not call them by name, since we often deal with special cases, where linearity is valid, and so we speak of representations of a group G. Nevertheless, the general case, where a group acts on a space in a nonlinear manner, appears quite often, and not only within the framework of Kaluza-Klein theory[2,3,4]. As an example I shall start with an application which concerns the standard model of elementary particle physics (see section 2).

G-spaces appear in a natural way whenever symmetries are present. We do not need to waste any words about the importance of symmetries in physics. The same is transferable to G-spaces, since these, in a sense, represent a geometrization of symmetry effects.

A G-space[1] is a space M on which a group (symmetry) acts in a "nice way". The group G should be, with respect to the internal symmetries, a compact Lie group and the space M a smooth manifold. That the group G acts in a "nice way" means that for every $g \in G$ the corresponding action f_g represents a diffeomorphism on M. Further, for $g, h \in G$ and $x \in M$ with

$$\begin{aligned} f : G \times M &\to M \\ (g,x) &\mapsto f(g,x) = g\cdot x = f_g(x) \ , \end{aligned}$$

$h(g\cdot x) = (h\cdot g)x$ and $1x = x$ should be valid. 1 is the neutral element of G. A manifold like this, equipped with an action of the group G, is called a G-manifold or, more generally, a G-space. In particular, one can have G-spaces with additional structure; we speak of G-fiber bundles, G-vector bundles...and even, in a corresponding generalization, of G-theories[5].

An especially simple example for a G-space is a global flow on a manifold M

$$f : \mathbb{R} \times M \to M \quad (\text{with } G := \mathbb{R}) .$$

The trajectory of a point $x \in M$ is an orbit $\mathbb{R}x$ of the $\mathbb{R}$-action. In this connection, the following question, with which we are going to deal later, is relevant: How many kinds of orbits are possible? The answer here is easy to find and quite simple: there are only three kinds of orbits;

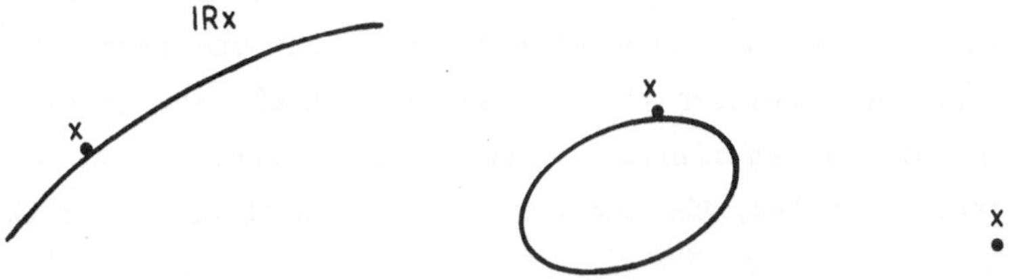

injective periodic fixed point

A further familiar example is the U(1) action on the vector space $\mathbb{R}^2$:

$$f : U(1) \times \mathbb{R}^2 \to \mathbb{R}^2$$

where

$$f_g = \begin{bmatrix} \cos\phi & -\sin\phi \\ \sin\phi & \cos\phi \end{bmatrix} .$$

We have two kinds of orbits here:

for $x \in \mathbb{R}^2$ and $x \neq 0$

$$U(1)\,\vec{x} \cong S^1$$

and

$$U(1)\,\vec{0} = \vec{0}.$$

The space of orbits $\mathbb{R}^2/U(1)$ is therefore the union of the zero point and the positive line $\mathbb{R}^+$. This is shown symbolically below.

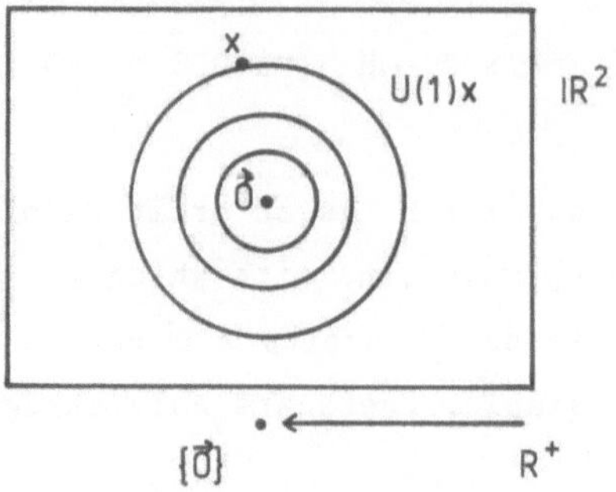

From these considerations you would probably guess that there must exist a lot of different kinds of G-spaces. But one should not overlook the fact that the condition that the group G should act (nontrivially) on M is a very strong requirement on the space M itself. As you also see from the above examples, the space M is composed of orbits of the group G.

In the following we shall discuss four different kinds of G-spaces. In section 2, the G-space $\mathring{M} = G/H$ with H a subgroup of G $(H \subset G)$ represents the ground state in the case of spontaneous symmetry breaking. It is composed of one orbit only. In section 3, the G-space $E = M \times G$ contains only one kind of orbit and every orbit is isomorphic to the group G. It represents a possible model for the Kaluza-Klein space. Section 4 corresponds to section 3 with the only difference that the orbits are isomorphic to the homogeneous space G/I $(I \subset G)$. In the last part, section 5, the G-space U, with a general G-action, is a kind of "pre-universe" from which the Kaluza-Klein space emerges, as in sections 3 and 4. One could represent these spaces symbolically as follows:

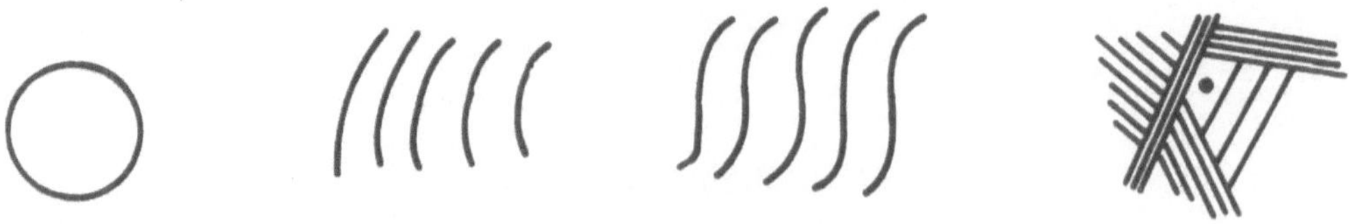

$\mathring{M}$ $\quad$ $E = M \tilde{\times} G$ $\quad$ $E = M \tilde{\times} G/I$ $\quad$ U

All these cases are closely related to a certain dynamic. But the dynamical aspects are kept in the background, since we mainly want to point out the geometrical aspects connected with G-spaces.

The ground state (vacuum) in a field theory with a symmetry G, which is spontaneously broken, is a very important example of a G-space. Since the situation here is well known from elementary particle physics, this example will be helpful to introduce and to explain some notions in connection with G-spaces. Spontaneous symmetry breaking is present whenever the ground state of the theory is degenerate. This is essentially described by the states of a spin zero field $\vec{\varphi}$ which minimizes the scalar potential V_φ. This is shown in the following picture:

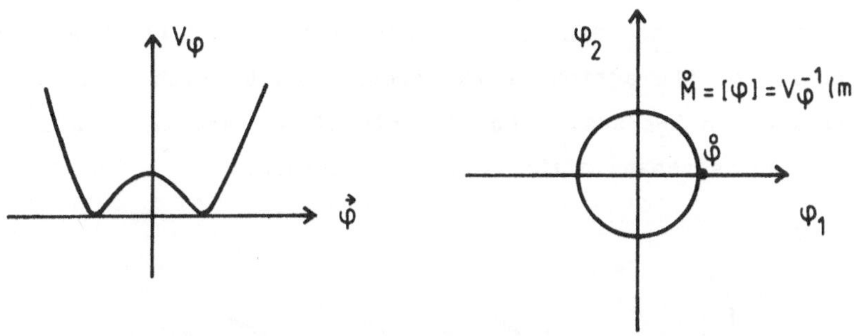

On the left-hand-side of the figure the graph of the scalar potential is shown, whereas on the right-hand-side the "φ-points in the φ-space" are shown, which correspond to the minimum of the potential. The bottom of a bottle of champagne would also be a suitable picture: the points of the bottom of the bottle build together a manifold $\overset{\circ}{M}$. Since all these points are equivalent, each one of them can be chosen as the actual vacuum, and therefore we are talking of the degeneracy of the ground state.

Because of the symmetry G existing in our theory, the group G acts also on the ground state $\overset{\circ}{M}$. So $\overset{\circ}{M}$ becomes a G-manifold. The following properties are especially relevant for our considerations:

(i) $\overset{\circ}{M}$ is the only orbit of the G-action (transitive action). This means that from a point $\overset{\circ}{\varphi} \in \overset{\circ}{M}$ the group G generates the whole space $\overset{\circ}{M}$ and we have $\overset{\circ}{M} = G\,\overset{\circ}{\varphi}$.

(ii) There exists a maximal subgroup H ($H \leq G$) so that $H\,\overset{\circ}{\varphi} = \overset{\circ}{\varphi}$. H is then the stability group of $\overset{\circ}{\varphi}$.

(iii) This H characterizes the orbit M. This means that the stability group H_1 of the elements $\varphi_1 \in \overset{\circ}{M}$ $(\varphi_1 \neq \overset{\circ}{\varphi})$ looks almost like the group H: We have from $H_1\varphi_1 = \varphi_1$ and $\varphi_1 = g\,\overset{\circ}{\varphi}$ (because of (i) this is possible with $g \in G$) $H_1 = g\,Hg^{-1}$. Since H and H_1 are of the same type, we can say that the orbit $\overset{\circ}{M}$ is of the orbit type (H).

(iv) There exists a model of $\overset{\circ}{M}$: $\overset{\circ}{M} \cong G/H$. This model produces the group G together with its subgroup H. The pair G,H contains all the secrets of $\overset{\circ}{M}$.

The homogeneous space G/H is the space of orbits of the H-action on a space G. It is now the action of the group H on G which makes out of G (in our notation) an H-space. The elements of G/H are the orbits of the H-action on G, as is shown in the following picture.

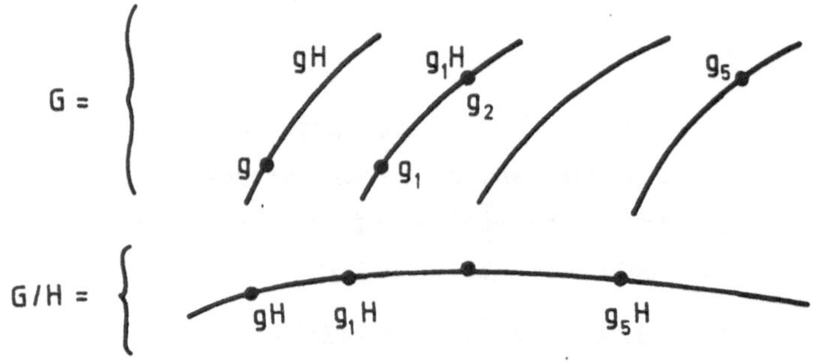

So we have $G/H = \{gH\} = \{[g]\}$. The equivalence relation $\sim$ which characterizes the elements of G/H is given by

$$g_1 \sim g_2 \quad \Leftrightarrow \quad g_1^{-1}g_2 \in H \quad \Leftrightarrow \quad g_2 = g_1h \text{ with } h \in H.$$

Now, G/H is itself a G-space, since we can define

$$G \times G/H \to G/H$$

$$(g, g_1H) \mapsto g(g_1H) = g\,g_1\,H.$$

With G/H we have now constructed a second G-space. This G-space is isomorphic to $\overset{\circ}{M}$ and we have, as mentioned, $\overset{\circ}{M} \cong G/H$. The mapping which describes this isomorphism is given by

$$i : G/H \to M$$
$$gH \mapsto g\,\overset{o}{\varphi}\,.$$

This mapping is well defined, injective, surjective and is compatible with the group action on M and H.

To illustrate the above concepts, we can now describe in an especially compact way the information we already have about spontaneous symmetry breaking. The indication of the orbit type characterizes the way the symmetry is spontaneously broken:

a) When (G) is the orbit type of $\overset{o}{M}$, then $\overset{o}{\varphi}$ is a fixpoint of G. In this case there is no symmetry breaking at all since $\overset{o}{M} \cong G/G = \{1\}$.

b) When (1) is the orbit type of $\overset{o}{M}$, then the action is free, so that $\overset{o}{M} \cong G/1 = G$ and the symmetry is totally broken.

c) When (I) is the orbit type of $\overset{o}{M}$, with $I < G$, then $\overset{o}{M} \cong G/I$ and the symmetry is broken down to the group I.

In the case of the chiral symmetries, for example, the case c) is realized. We have (for two flavours)

$$SU(2)_L \times SU(2)_R = G$$
$$SU(2)_V = I$$

and the ground state is isomorphic to the sphere S^3 given by

$$\frac{SU(2) \times SU(2)}{SU(2)} = G/I.$$

3. THE KALUZA-KLEIN THEORY 1: THE GROUP ACTS SIMPLY AND FREELY

3.1) Some Very General Considerations

The G-space E we would like to consider now is slightly more complicated than the previous one. E consists of many orbits and all of them belong to the same orbit type (1).

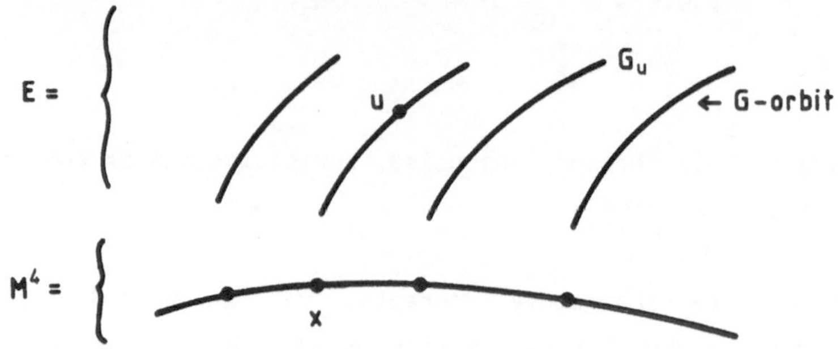

Every orbit is isomorphic to the group G. In this case we can say that the group G acts freely (but not transitively) on E. "Freely" means that no subgroup of G, except the trivial one $\{1\}$, is allowed to have fixpoints. "Simply" means that only one orbit type is present. So "simply and freely" is in this case obvious and is stated only to be compared with the next section. From the fact that the action is not transitive, it follows that the space of orbits is nontrivial and we have

$$M = E/G = \{uG\} \quad \text{with } u \in E ,$$

so that the Kaluza-Klein space E has, at least locally, the structure

$$E = M \tilde{\times} G.$$

It is a principal bundle with structure group G. (We use the symbol $\tilde{\times}$ to indicate the local character of the product).

It is perhaps useful, at this point, to pause and remember how we arrived at considerations of this kind within the framework of elementary particle physics. The fact that all fundamental interactions in nature have as a central characteristic the gauge-structure brought elementary particle physics and gravitation very close together, and the call for unified theory again became timely. This is the requirement which Einstein always put forward with great emphasis. It is notable that the opinion that it must be possible to describe nature within a unified theory was also a deep philosophical insight of the pre-Socratic philosophers. The Pythagoreans, especially after the $\sqrt{2}$ shock[6], also held the opinion that nature should be described by a geometrical theory. As we see, "geometrical trends", the subject of this School, is not only a very modern but also a truly modern theme. For what is more modern than an eternal truth? The important contribution of Kaluza in this connection was that he proposed a certain realization, a five-dimensional world, for the unification of gravitation and electrodynamics.

Realizing the geometric character of all interactions, new and unavoidable questions arise.

- How many dimensions does our universe have?
- What is its global topological structure?
- Are inner quantum numbers of elementary particles capable of revealing secrets of the space-time structure?

Is the latter question not perhaps exaggerated? The history of electrodynamics and general relativity suggests that it is not. After the discovery of Riemannian geometry efforts were made by Riemann himself, and probably by many others, to understand nature within the framework of Riemannian geometry. But the discovery of the general theory of relativity at that period was not possible, since the nature of time was not clarified. It was electrodynamics, together with its inner quantum numbers (the electric charge), which allowed an understanding of the role that time plays in modern physics, and this in turn lead immediately to the Minkowskian space-time. After that Einstein, relying on Minkowski's space-time and on the equivalence principle, discovered general relativity. So it is not completely absurd to suppose that inner quantum numbers have something to teach us about the nature of space-time.

3.2) On the Kaluza-Klein idea

A slight generalization of the Kaluza-Klein idea is the concept that our space-time M^4 should be imbedded in a bigger universe E. Let $E = M \tilde{\times} S$, where S is a compact manifold representing certain inner quantum numbers. Two main questions immediately arise:

i) How does such a universe come about?

ii) What follows from this idea for our space-time?

These two questions are often characterized by the notions of spontaneous compactification and reduction, respectively. Here we shall deal with the second question and in section 5 we shall return to the first one.

So we start with a universe E which has locally the structure M x S and ask ourselves first what properties E must have in order to avoid the observability of the additional dimensions when we are doing macroscopic experiments on M^4 and, in addition, what follows for our space-time. There exist a lot of contributions to this problem[2,3,4,7]; for a recent review, even written in Kaluza's language, see Ref.[8].

Within the geometrical framework it is possible to give a very compact and precise formulation of this problem, and subsequently its solution, which I would like to call the Kaluza-Klein theorem[7]:

> If on the Kaluza-Klein space E the group G acts freely in such a way that it leaves the metric g_E invariant, general relativity on E is equivalent to the following physics on M^4:

i) A general relativity on $M^4 = E/G$ given by the metric $g_{\mu\nu}$ on M.

ii) A Yang-Mills theory on M^4 with the structure group G given by the gauge potentials A^j_μ ; the index j belongs to the Lie algebra of G.

iii) A kind of σ-model on M^4, given by the scalar fields g_{ij} which describe the invariant metric on G.

All these fields on M ($g_{\mu\nu}$, A^j_μ and g_{ij}) appear in the metric g_E on E and in this way we can summarize the above theorem, in a so-called direct product basis, by the following representation:

$$g_E = \left[\begin{array}{c|c} g_{\mu\nu} + A^j_\mu A^k_\nu g_{jk} & A^i_\nu g_{ik} \\ \hline A^i_\mu g_{ij} & g_{jk} \end{array}\right]$$

3.3) On the Interpretation of the Covariant Derivative

As we have seen, the metric g_E on the Kaluza-Klein space E contains the Yang-Mills potentials A^j_μ in a very special way. This allows an interpretation of the covariant derivative as a special horizontal vector field on E. The following considerations will also be useful in the next section, when we are looking for the structure group of the Yang-Mills theory in a more general case than those previously considered.

On the tangential space TE of E, because of the G-action, we have n ($n = \dim G$) linear independent vector fields ${}^{\supset}T_j$ with $j = 1,2,\dots,n$ which are tangential to the orbits. These ${}^{\supset}T_j$'s are generated by the one-parameter subgroups which correspond to the basis T_j of the Lie algebra of G. They span at every point $u \in E$ a subspace of T_uE with dimension n, and we call it the vertical space: ver T_uE. The metric g_E allows us to determine its orthogonal complement (ver $T_uE)^{\perp}$. This is called the horizonal space, hor T_uE = (ver $T_uE)^{\perp}$, and has of course the same dimension as M, since we have

$$T_uE = \text{ver } T_uE \oplus (\text{ver } T_uE)^{\perp} .$$

In particular, hor $T_uE \cong T_xM$ is valid, where $x \in M$ is the projection of the point u.

We now consider the local imbedding $\bar{M}$ of M in E. Through this imbedding the point x goes to u, and accordingly the vector field ∂_μ on M to the vector field $\bar{\partial}_\mu$ on E, as is shown in the figure:

The vector $\bar{\partial}_\mu(u) =: \bar{\partial}_\mu$ is at a general position in T_uE and we denote by ver $\bar{\partial}_\mu$ and hor $\bar{\partial}_\mu = D_\mu$ its projections in the vertical and horizontal spaces, as in the following figure:

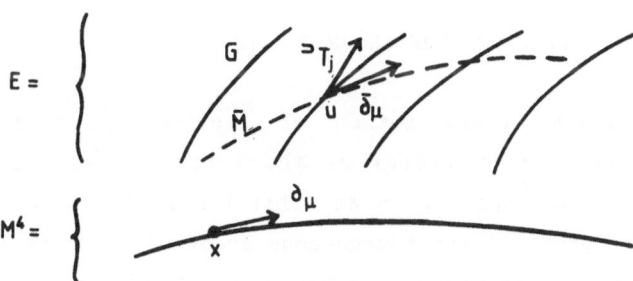

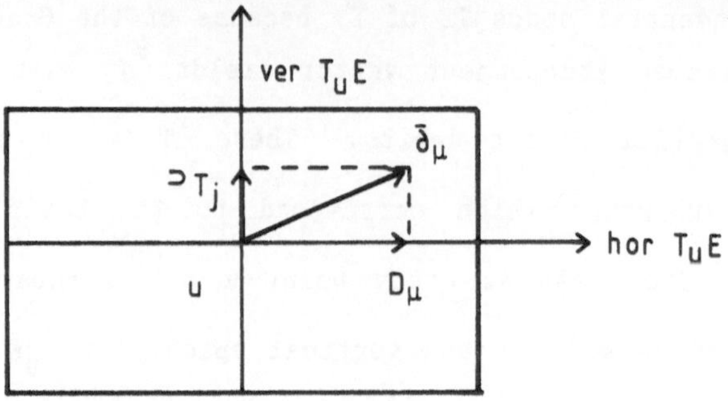

Now, as the vector $^{\supset}T_j$ generates the vector space, we can write

$$\text{ver}\ \bar{\partial}_\mu = A^j_\mu\ {}^{\supset}T_j\ .$$

So we have for the horizontal projection of $\bar{\partial}_\mu$

$$D_\mu = \bar{\partial}_\mu - A^j_\mu\ {}^{\supset}T_j\ .$$

After a plausible identification we obtain the usual form of the covariant derivative

$$D_\mu = \partial_\mu - A^j_\mu\ T_j\ .$$

As we see, the local imbedding of M in E has, by the given structure, revealed the covariant derivative.

4. KALUZA-KLEIN THEORY 2: THE GROUP G ACTS SIMPLY

As a slight generalization of section 3, we now regard the Kaluza-Klein space with orbits of the type (I), with I a non-trivial subgroup of G. The space E is an orbit bundle (OB) but its fibres are, instead of the group G, the homogeneous space G/I. Because of the strong similarity to the previous section I shall not go into detail, but concentrate on the question: what should the structure group of the corresponding Yang-Mills theory on M^4 be in this case?

Spontaneously, one would think that the group G is the structure group, or some sceptics would probably prefer the subgroup I. It is perhaps interesting to say, anticipating, that in general neither the spontaneous nor the sceptical answer would be right.

On the basis of the considerations in section 3.3 we can give an intuitive derivation of the structure group of the Yang-Mills theory. (The exact derivation is given in Ref. [7].) The point is, in fact, that we are looking for the Lie algebra components of the imbedded vector $\partial_\mu \in T_u E$. Or, analogously, we would like to know, pictorially speaking, into what group K M locally goes.

The space G/I is in general not a group. In order to obtain a group in connection with I in the quotient space G/I we could use the normalizer N(I) of I in G. N(I) is so defined that N(I)/I is a group. So we have for every orbit

$$G/I = G/N \tilde{\times} N/I = G/N \tilde{\times} K \ .$$

Now we can locally regard the subbundle $M \tilde{\times} K$ of E. To this we can apply all the considerations of section 3.3. So we see immediately that K is the structure group of the Yang-Mills theory on M^4.

5. KALUZA-KLEIN THEORY 3: U WITH A GENERAL G-ACTION

5.1 The Problem

Until now we have started with general relativity on a very special G-space (consisting of only one orbit bundle). Then, by reduction (Kaluza-Klein theorem), it was possible to obtain the corresponding theory in four dimensions. Here we would like to make a very drastic generalization[5]. This is of particular importance in connection with the question of spontaneous compactification.

We start with a universe U on which general relativity is valid. U should in fact be a G-space, but with essentially no further constraints. The question now is whether it is possible to obtain something like the Kaluza-Klein space as in sections 3 and 4, having started with such a general structure. We expect, indeed, that U will look very complicated, and in general will contain numerous orbit types with various dimensions, e.g. $(I_1),(I_2),\ldots,(I_k),(H)$, where $I_1,\ldots,I_k,H$ are subgroups of G. In the last figure of section 1 such a U was shown symbolically. As we see, the main difference to sections 3 and 4 is that there the space E had only one orbit type (I).

Despite the fact that the situation with U looks quite chaotic, there exist some methods from equivariant differential topology[1] which allow us to deal with the problem. In partucular, the slice theorem is a powerful instrument to investigate the interplay of the various orbit types.

5.2 The Slice Theorem

This theorem gives information about the neighbourhood of an orbit. As we showed in section 2, the group G itself can produce the model of an orbit as soon as we know the stability group I in one of its points $u \in U$. So we have

$$Gu \cong G/I .$$

The slice theorem is a generalization of this, as it also gives a model of the neighbourhood of an orbit. The only information we need in addition is how the isotropy group I acts on the tangential space at this point (T_uU). The essential content of the theorem is the following: There exists a neighbourhood of the orbit Gu which is isomorphic to the vector bundle $\mathscr{V}_{(I)}$, which has the space G/I as its base manifold and the vector space V (a subspace of T_uU) as a typical fibre. The vector space V itself is a representation space of the group I. So in summary we have

$$Nb(Gu) \cong G \underset{I}{x} V$$

where Nb represents the neighbourhood and $G \underset{I}{x} V = \mathscr{V}_{(I)}$. $G \underset{I}{x} V$ gives precise local and global information on the vector bundle $\mathscr{V}_{(I)}$.

To understand the neighbourhood of the orbit Gu means to understand this vector bundle; this is what we would like to explain next. I in the above notation means the structure group of the vector bundle and G that $\mathscr{V}$ is an associated bundle to the principal bundle G(G/I,I). So we can write

$$\mathscr{V}_{(I)} = G/I \tilde{x} V .$$

What is left now is to specify the vector space V. Using the fact that I is the stability group of u we have, for $h \in I$, $hu = u$ and therefore

$$dh(u) : T_uU \to T_uU .$$

Since $dh(U)$ operates on T_uGu like the identity, the normal space $(T_uU)_{nor} := T_uU/T_u(Gu)$ is a representation space of I. V is precisely this normal space, and the corresponding representation is usually called the slice representation. Taking this vector space, we can construct the vector bundle $G \underset{I}{x} V = \mathscr{V}_{(I)}$ using the information given by the principal bundle G(G/I,I). The result is represented symbolically in the next figure.

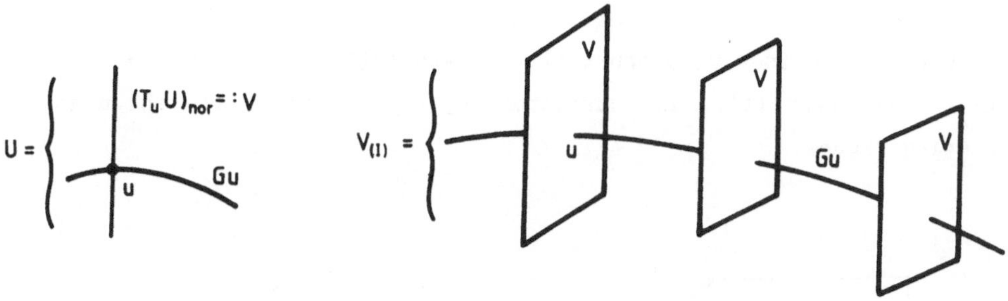

It is interesting to note that V is an I-space and that the orbits, together with the corresponding I orbit types, characterize completely the neighbourhood of the whole G orbit on U. The I-space is symbolically represented in the following figure:

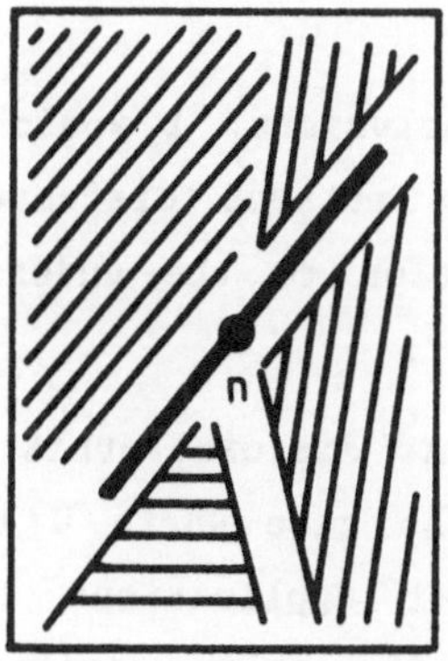

V : I-space

In order to give an especially simple example for the slice theorem, we regard the sphere S^2 together with the U(1) action, as shown in the figure:

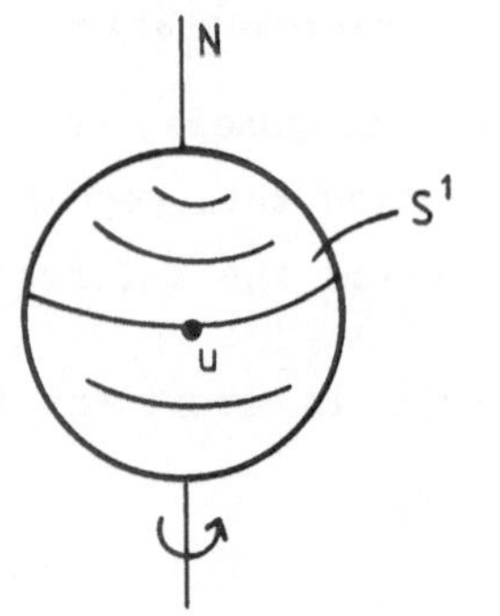

The North Pole N is a U(1) orbit of the type (U(1)). The equator S^1 is an orbit of the type (1). The corresponding slice bundles are shown in the following figure:

The slice bundle for N is a two-dimensional vector space, the slice bundle for the equator is an $\mathbb{R}^1$ vector bundle with S^1 as its base manifold.

5.3. Some Results

Finally, instead of conclusions, I would like to give some results which were obtained by the methods discussed above, and which could furnish a certain contribution to the understanding of the problem of spontaneous compactification :

(i) The universe splits into disjoint orbit bundles, every one of which is a manifold. In the case where U is compact, their number is finite. For physical applications this finiteness has to be demanded, and it is the only condition we have to put on the G-action on U.

(ii) Every orbit bundle is a fibre bundle with typical fibre a homogeneous space, as in sections 3 and 4, and the structure $M_j \times G/I_j$.

(iii) There always exists exactly one orbit bundle E (of orbit type (H), $H \leq G$) which has the same dimension as the universe U, so we have dim E = dim U; all the other orbit bundles have a lower dimension.

(iv) With the corresponding stability group we obtain

$$\begin{array}{ccccccc} U = E & \dot{\cup} & U_{(1)} & \dot{\cup} & \cdots & \dot{\cup} & U_{(k)} \\ \downarrow & & \downarrow & & & & \downarrow \\ M^4 & \dot{\cup} & M_{(1)} & \dot{\cup} & \cdots & \dot{\cup} & M_{(k)} \end{array}$$

(v) The orbit bundle E (principle orbit bundle) constitutes our Kaluza-Klein space, and we can apply the reduction formalism as in sections 3 and 4.

(vi) The remaining orbit bundles of lower dimension might have a physical interpretation and in any case have to be taken into account [9].

ACKNOWLEDGEMENT

I would like to thank the organizers, the participants and the personnel of the Physik-Zentrum Bad Honnef for the excellent atmosphere at the School.

REFERENCES

1. G. E. Bredon, "Introduction to Compact Transformation Groups", Academic Press, New York (1972);
K. Jähnich, Lecture Notes in Mathematics 59, Springer, Berlin and Heidelberg (1968).
2. Y. M. Cho, J. Math. Phys. 16: 2029 (1975).
3. Th. Kaluza, Sitzungsber. Preuss. Akad. d. Wiss. Berlin, Math. Phys. Klasse: 966 (1921);
O. Klein, Z. Phys. 37: 895 (1926);
A. Einstein und P. Bergmann, Ann. Math. 39: 683 (1938);
J. M. Souriau, Nuovo Cim. 30: 565 (1963);

A. Trautmann, Rep. Math. Phys. 1: 29 (1970);

J. Raayski, Acta Phys. Pol. 27: 89 (1965)

4. See e.g. M. J. Duff, B. E. W. Nilsson and C. N. Pope, Phys. Rep. 130: 1 (1986) and references therein.

5. A. Heil, B. Reifenhäuser, N. A. Papadopoulos and F. Scheck, "G-Theories and the geometric compactification scheme", Nucl. Phys. B, to appear.

6. K. R. Popper (E. Papadaki), private communication.

7. R. Coquereaux and A. Jadczyk, Comm. Math. Phys. 90: 79 (1983);

R. Coquereaux and A. Jadczyk, Comm. Math. Phys. 98: 79 (1985).

8. E. W. Mielke, Jahrbuch Überblicke Mathematik, Bibliographisches Institut, (1986).

9. A. Heil, N. A. Papadopoulos, B. Reifenhäuser and F. Scheck, Phys. Lett. B173: 517 (1986);

A. Heil, N. A. Papadopoulos, B. Reifenhäuser and F. Scheck, Nucl. Phys. B281: 42 (1987).

II. ANOMALIES

THE ORIGIN AND SIGNIFICIANCE OF CHIRAL AND GRAVITATIONAL ANOMALIES

H. Leutwyler

GAUGE INVARIANCE IN SPITE OF ANOMALIES

N. K. Falck

ANOMALIES IN ODD DIMENSIONS

M. Reuter

THE ORIGIN AND SIGNIFICANCE
OF CHIRAL AND GRAVITATIONAL ANOMALIES

H. Leutwyler

Institute for Theoretical Physics, University of Bern
Sidlerstr. 5, CH-3012 Bern, Switzerland

ABSTRACT

The lecture reviews the origin and the significance of chiral and gravitational anomalies. The phenomenon is analyzed in terms of the short distance singularities generated by free fermion fields, using two-dimensional space-time as a guide.

INTRODUCTION

One of the basic features of the standard model is the fact that it contains gauge fields which only interact with left-handed fermions. The boson mediating the charged weak interaction, e.g., couples to the fermions through a current of the type V-A. It is of crucial importance for the consistency of the model that the gauge fields couple to conserved currents. The short distance singularities of the fermion fields in general, however, spoil the conservation of the axial currents: the conservation laws are afflicted with anomalies[1], which ruin gauge invariance. The standard perturbative analysis of gauge field theories with left-handed couplings is consistent only if these anomalies happen to cancel[2].

The first part of this review deals with free fermions. The Ward identities for the free currents contain anomalous contributions which originate in the short distance singularities of the Fermi fields. The implications of the free field analysis for gauge field theory are addressed in the second part of the talk. The third part concerns gravitational anomalies. Global anomalies[3,4] and topological aspects[4-7] are not covered.

1. GREEN'S FUNCTIONS OF THE FREE VECTOR CURRENT

As announced above, I first consider a free Dirac field, for simplicity taken to be massless

$$-i\gamma^{\mu}\partial_{\mu}\psi(x) = 0 \ . \tag{1}$$

The corresponding vector and axial currents

$$V^{\mu} = :\bar{\psi}\gamma^{\mu}\psi: \ ; \qquad A^{\mu} = :\bar{\psi}\gamma^{\mu}\gamma_5\psi: \tag{2}$$

and the left-handed current

$$L^{\mu} = :\bar{\psi}\gamma^{\mu}\tfrac{1}{2}(1-\gamma_5)\psi:= \tfrac{1}{2}(V^{\mu}-A^{\mu}) \tag{3}$$

are conserved

$$\partial_{\mu}V^{\mu} = \partial_{\mu}A^{\mu} = \partial_{\mu}L^{\mu} = 0 \ . \tag{4}$$

Let us first look at the Green's functions of the vector current. The two-point-function, e.g. is given by

$$G^{\mu\nu}(x-y) = \langle 0|TV^{\mu}(x)V^{\nu}(y)|0\rangle \tag{5}$$

$$= \mathrm{tr}\,\{\gamma^{\mu}S_0(x-y)\gamma^{\nu}S_0(y-x)\}\ ,$$

where $S_0(x-y)$ is the free propagator

$$S_0(z) = -\frac{z_{\mu}\gamma^{\mu}}{2\pi^2(z^2 - i\epsilon)^2}\ . \tag{6}$$

The distribution $G^{\mu\nu}(z)$ is well defined only on test functions which vanish at $z = 0$, together with their first and second derivatives. On these, it obeys the Ward identity

$$\partial_{\mu}G^{\mu\nu}(z) = 0\ . \tag{7}$$

The extension of $G^{\mu\nu}(z)$ to arbitrary test functions is not unique. There are different extensions which obey (7) on all test functions and are Lorentz invariant. In fact, these two requirements determine $G^{\mu\nu}(z)$ up to one free parameter: two Lorentz invariant extensions which obey (7) differ by

$$\overline{G}^{\mu\nu}(z) = G^{\mu\nu}(z) + ic(g^{\mu\nu}\Box - \partial^{\mu}\partial^{\nu})\delta(z)\ . \tag{8}$$

The three - point - function $\langle 0|TV^{\lambda}V^{\mu}V^{\nu}|0\rangle$ vanishes on account of charge conjugation invariance. The four - point - function $\langle 0|TV^{\mu_1}(x_1)\ldots V^{\mu_4}(x_4)|0\rangle_{\mathrm{conn}}$ is unique up to a local term proportional to $\delta(x_1-x_2)\delta(x_1-x_3)\delta(x_1-x_4)$ (logarithmic divergence in the corresponding box graph). In fact, current conservation fixes the extension completely. Finally, Green's functions with more than 4 currents are unambiguous and are conserved. To summarize:

(i) The Green's functions of the free vector current can be chosen such that

$$\partial_{\mu_1}\langle 0|TV^{\mu_1}V^{\mu_2}\ldots V^{\mu_n}|0\rangle = 0\ . \tag{9}$$

(ii) There is a renormalization ambiguity only in the two-point-function.

2. GENERATING FUNCTIONAL

It is convenient to collect all Green's functions of the vector current in the generating functional $Z_V(f)$, defined by

$$e^{iZ_V(f)} = \langle 0|Te^{i\int dx f_\mu(x)V^\mu(x)}|0\rangle . \tag{10}$$

The individual connected Green's functions are obtained from $Z_V(f)$ by expanding in powers of the external field $f_\mu(x)$:

$$\begin{aligned} Z_V(f) = & \frac{i}{2!}\int dx_1 dx_2 f_{\mu_1}(x_1) f_{\mu_2}(x_2) \langle 0|TV^{\mu_1}(x_1)V^{\mu_2}(x_2)|0\rangle \\ & + \frac{i^3}{4!}\int dx_1 \ldots dx_4 f_{\mu_1}(x_1)\ldots f_{\mu_4}(x_4)\langle 0|TV^{\mu_1}(x_1)\ldots V^{\mu_4}(x_4)|0\rangle_{conn} \\ & + \ldots \end{aligned}$$

The Ward identities (9) amount to the statement that the generating functional is invariant under the gauge transformation

$$f'_\mu(x) = f_\mu(x) + \partial_\mu \alpha(x) \; ; \quad Z_V(f + \partial\alpha) = Z_V(f) . \tag{11}$$

The fact that only the two-point-function contains a renormalization ambiguity, given by (8), is equivalent to the statement that the generating functional is unique up to a local polynomial in the external field:

$$\bar{Z}_V(f) = Z_V(f) + \frac{c}{4}\int dx \, (\partial_\mu f_\nu - \partial_\nu f_\mu)^2 . \tag{12}$$

3. GREEN'S FUNCTIONS OF THE FREE LEFT-HANDED CURRENT

Consider now the generating functional associated with the Green's functions of the left-handed current

$$e^{iZ_L(f)} = \langle 0|Te^{i\int dx f_\mu(x)L^\mu(x)}|0\rangle . \tag{13}$$

Again, the short distance singularities of the free propagator generate ambiguities in the Green's functions containing up to four currents. It is possible to renormalize the two- and the four-point-functions in such a manner that current conservation holds. In the case of the three-point-function this turns out to be impossible: independently of how one extends the distribution $\langle 0|TL^{\lambda}(x)L^{\mu}(y)L^{\nu}(z)|0\rangle$ to the space of all test functions, at least one of the three currents fails to be conserved at $x = y = z$. If the three-point function is renormalized in such a manner that it is symmetric with respect to the interchange of any two of the three currents, one finds

$$\partial_{\lambda}\langle 0|TL^{\lambda}(x)L^{\mu}(y)L^{\nu}(z)|0\rangle = -\frac{1}{12\pi^2}\,\epsilon^{\mu\nu\alpha\beta}\,\partial^{y}_{\alpha}\,\partial^{z}_{\beta}\,\{\delta(x-y)\delta(x-z)\}\ .$$

Accordingly, the generating functional is not gauge invariant:

$$Z_L(f+\partial\alpha) = Z_L(f) + \frac{1}{24\pi^2}\int dx\ \alpha(x)\ \epsilon^{\mu\nu\rho\sigma}\ \partial_{\mu}f_{\nu}(x)\ \partial_{\rho}f_{\sigma}(x)\ . \qquad (14)$$

To see how the phenomenon arises, let us consider the left-handed current in two-dimensional space-time. With $\gamma_5 = \gamma_0\gamma_1$ the vector and axial currents are related by $A^0 = -V^1$, $A^1 = -V^0$. The component $L^- = L^0 - L^1$ of the left-handed current therefore vanishes identically, $L^-(x) = 0$. Furthermore, current conservation implies that $L^+ = L^0 + L^1$ is a free field, depending only on $x^0 - x^1$;

$$\partial_+ L^+(x) = 0\ . \qquad (15)$$

The free propagator is

$$S_0(z) = \frac{z^{\mu}\gamma_{\mu}}{2\pi(z^2 - i\epsilon)} \qquad (16)$$

and the two-point-function therefore becomes

$$\langle 0|TL^+(x)L^+(y)|0\rangle = -tr(\gamma^+ \not{z}\gamma^+ \not{z})/4\pi^2(z^2-i\epsilon)^2 = \frac{i}{\pi}\,\partial_-\partial_-\Delta(z)\ , \qquad (17)$$

where $z = x-y$ and where $\Delta(z)$ is the scalar propagator

$$\Delta(z) = \frac{1}{4\pi i}\ln(-z^2+i\epsilon)\ ; \qquad \Box\Delta(z) = \delta(z)\ . \qquad (18)$$

In two dimensions, the Ward identity for the two-point-function therefore contains an anomaly:

$$\partial_+\langle 0|TL^+(x)L^+(y)|0\rangle = \frac{i}{\pi}\,\partial_-\delta(z)\ . \qquad (19)$$

In $d=2$ all connected Green's functions containing more than two currents vanish ($L^+(x)$ is a free field). The generating functional can therefore be given in closed form[8]:

$$Z_L(f) = \frac{1}{8\pi} \int dx\, dy\, \partial_- f_+(x)\, \Delta(x-y)\, \partial_- f_+(y) . \tag{20}$$

Obviously, this expression is not gauge invariant.

If the dimension d of space-time is odd, there is no analogue of γ_5 (the product $\gamma_0\gamma_1 \ldots \gamma_{d-1}$ is a multiple of the unit matrix); there are no left-handed or axial currents and there are no anomalies.

If d is even, there is an anomaly in the Ward identities for Green's functions with $d/2+1$ left-handed currents (vacuum polarization diagram in d=2, triangle graph in d=4, hexagon diagram in d=10 etc.)

4. FERMIONS IN AN EXTERNAL FIELD

The generating functional admits an alternative interpretation: it represents the vacuum-to-vacuum transition amplitude in the presence of an external field. To see this, consider the Lagrangian

$$\mathcal{L} = i\chi^+ \tilde{\sigma}^\mu (\partial_\mu - i f_\mu)\, \chi , \tag{21}$$

where $\chi(x)$ is a two-component spinor describing left-handed fermions and where $f_\mu(x)$ is an external field. The 2×2 Weyl matrices σ^μ, $\tilde{\sigma}^\mu$ represent the right- and the left-handed components of the Dirac matrices

$$\gamma_\mu = \begin{bmatrix} 0 & \tilde{\sigma}_\mu \\ \sigma_\mu & 0 \end{bmatrix} ; \quad \gamma_5 = \begin{bmatrix} 1 & 0 \\ 0 & -1 \end{bmatrix} . \tag{22}$$

If the external field vanishes for $x^0 \to \pm\infty$, the field $\chi(x)$ develops from a free incoming field $\chi_{in}(x)$ to a free outgoing field $\chi_{out}(x)$. Denote the ground states of these free fields by $|0\ in\rangle$ and $|0\ out\rangle$ respectively. The probability for the external field not to create fermion pairs is given by $|\langle 0\ out|0\ in\rangle|^2$, where

$$\langle 0\ out|0\ in\rangle = \langle 0\ in|T e^{i\int dx \mathcal{L}_{int}}|0\ in\rangle = \langle 0\ in|T e^{i\int dx f_\mu \chi_{in}^+ \tilde{\sigma}^\mu \chi_{in}}|0\ in\rangle . \tag{23}$$

The operator $\chi_{in}^{+} \tilde{\sigma} \chi_{in}$ is the left-handed current L^{μ} associated with the incoming free field. The vacuum-to-vacuum transition amplitude is therefore given by

$$\langle 0 \text{ out} | 0 \text{ in} \rangle = e^{iZ_L(f)} , \tag{24}$$

where $Z_L(f)$ is the generating functional of the free left-handed current discussed above. The presence of an anomaly in the free left-handed current therefore implies that the vacuum-to-vacuum transition amplitude is not gauge invariant:

$$\langle 0 \text{ out} | 0 \text{ in} \rangle_{f+\partial\alpha} = \langle 0 \text{ out} | 0 \text{ in} \rangle_{f} \, e^{i/24\pi^2 \int dx \alpha \epsilon^{\mu\nu\rho\sigma} \partial_{\mu} f_{\mu} \partial_{\rho} f_{\sigma}} . \tag{25}$$

The anomaly only affects the phase of the transition amplitude; the transition probability is gauge invariant.

5. DETERMINANT

The vacuum-to-vacuum transition amplitude can be represented as a fermionic path integral

$$\langle 0 \text{ out} | 0 \text{ in} \rangle = \int [d\chi] e^{-i\int dx \; \chi^{+} D_L \chi} ,$$

where D_L denotes the Weyl operator

$$D_L = -i\tilde{\sigma}^{\mu}(\partial_{\mu} - if_{\mu}) . \tag{26}$$

Since the exponential is quadratic in the variables of integration, the integral can be done:

$$\langle 0 \text{ out} | 0 \text{ in} \rangle = \det D_L . \tag{27}$$

To check that the determinant of the Weyl operator indeed coincides with $\exp iZ_L(f)$, we expand it in powers of the external field:

$$D_L = D_L^o - f \ ; \quad f = \tilde{\sigma}^\mu f_\mu$$

$$\ln \det D_L = \ln \det D_L^o - \mathrm{Tr}(fS_o) - \tfrac{1}{2}\,\mathrm{Tr}(fS_ofS_o) - \ldots$$

The symbol $S_o = (D_L^o)^{-1}$ denotes the free propagator. If we normalize the determinant of the free Weyl operator to $\det D_L^o = 1$, the series expansion for $\ln \det D_L$ is identical with the series expansion for $iZ_L(f)$. In particular, the renormalization ambiguities in $\ln \det D_L$ are the same as in the generating functional: the logarithm of the Weyl determinant is well defined up to a local polynomial in the external field.

To give unambiguous meaning to the determinant of the Weyl operator, we use Schwinger's method[9]: instead of specifying the determinant directly, we specify the change in $\det D_L$ produced by a change in the external field. Formally, the change is given by

$$\delta \ln \det D_L = \mathrm{Tr}(\delta D_L D_L{}^{-1}) = - \int dx\ \delta f_\mu(x)\ \mathrm{tr}\{\ \tilde{\sigma}^\mu S_f(x,x)\ \}\ , \tag{28}$$

where $S_f(x,y)$ is the propagator in an external field

$$-i\tilde{\sigma}^\mu\ (\partial_\mu - if_\mu(x))\ S_f(x,y) = \delta(x-y)\ . \tag{29}$$

Since the propagator is singular at $x = y$, the quantity $S_f(x,x)$ does not make sense as it stands. In $d=2$, e.g., the short distance behaviour of $S_f(x,y)$ is

$$S_f(x,y) = \frac{z^\mu \sigma_\mu}{2\pi(z^2 - i\epsilon)}\ \{1 + iz^\nu f_\nu(x)\} + \hat{S}_f(x,y)\ , \tag{30}$$

where the remainder, $\hat{S}_f(x,y)$, approaches a well defined distribution $\hat{S}_f(x,x)$ as $x \to y$. The essential point here is that terms which explode as $x \to y$ are local polynomials in the external field,

$$S_f(x,y) = P_f(x,y) + \hat{S}_f(x,y)\ . \tag{31}$$

In $d=2$, the polynomial $P_f(x,y)$ is linear in $f_\mu(x)$, in $d>2$ the singular part involves derivatives of f_μ as well as higher powers of f_μ. To give meaning to (28), we remove the singular terms and specify the change of the renormalized determinant as

$$\delta \ln \det D_L = - \int dx\, \delta f_\mu(x)\, \mathrm{tr}\{\tilde{\sigma}^\mu\, \hat{S}_f(x,x)\} \;. \qquad (32)$$

If one considers arbitrary deformations of the external field, there is an integrability problem with this formula. We avoid this problem by considering only a special class of deformations which interpolate the external field $f_\mu(x)$ of interest with $f_\mu = 0$:

$$f^t_\mu(x) = t f_\mu(x) \;; \qquad \delta f_\mu(x) = \delta t f_\mu(x) \;. \qquad (33)$$

Integrating the changes produced in $\ln \det D_L$ as the parameter t is increased from 0 to 1, we get

$$\ln \det D_L = - \int_0^1 dt \int dx\, f_\mu(x)\, \mathrm{tr}\, \{\tilde{\sigma}^\mu\, \hat{S}_{f^t}(x,x)\} \;. \qquad (34)$$

This formula provides us with an explicit expression for the generating functional in terms of the external field propagator $S_f(x,y)$. To analyze the properties of the generating functional it therefore suffices to analyze the properties of the differential equation which determines the propagator.

As emphasized above, $\ln \det D_L$ is well defined only up to a local polynomial. In the representation (34) of the determinant, this ambiguity shows up in the fact that the finite part $\hat{S}_f(x,x)$ of the propagator is not unique, because only the singular part of the polynomial $P_f(x,y)$ is unambiguous. In $d=2$, e.g., the polynomial given in (30) can be replaced by

$$\overline{P}_f(x,y) = P_f(x,y) + c\, \sigma^\mu f_\mu(x) \;. \qquad (35)$$

This modification changes $\ln \det D_L$ by a local polynomial:

$$\ln \overline{\det D_L} = \ln \det D_L + c \int dx\, f_\mu(x) f^\mu(x) \;. \qquad (36)$$

In fact, the explicit expression (20) for the generating functional in two dimensions differs from the determinant specified in (30), (34) by such a local polynomial:

$$iZ_L(f) = \ln \det D_L - \frac{i}{8\pi}\int dx\, f_\mu f^\mu . \tag{37}$$

In two dimensions, the Weyl operator D_L only contains the component $f_+(x)$ of the external field - left-handed fermions do not interact with $f_-(x)$. This property is borne out in the generating functional $Z_L(f)$, but it is not borne out in the above renormalization of the Weyl determinant which does depend on $f_-(x)$ through the local polynomial $f_\mu f^\mu = f_+f_-$. It is clear where this comes from: the choice (30) of the local polynomial $P_f(x,y)$ involves a finite piece proportional to $f_-(x)$. The renormalization prescription (34) therefore contains a finite 'counter term' which depends on f_-. This is artificial - the singular part of $S_f(x,y)$ only contains f_+, because the Weyl operator only contains f_+.

The origin of the anomaly can easily be seen in the above representation for the generating functional. Under a gauge transformation, the Weyl operator transforms according to

$$f'_\mu = f_\mu + \partial_\mu\alpha \; ; \qquad D'_L = e^{i\alpha} D_L e^{-i\alpha} . \tag{38}$$

The propagator $S_f(x,y)$ transforms in the same fashion; the polynomial $P_f(x,y)$ can be chosen such that the finite part $\hat{S}_f(x,x)$ is gauge invariant. We therefore obtain

$$\ln \det D'_L = \ln \det D_L - \int_1^0 dt \int dx\, \partial_\mu\alpha\, \mathrm{tr}\{\tilde{\sigma}^\mu \hat{S}_{f^t}(x,x)\} . \tag{39}$$

To evaluate the right-hand side we integrate by parts. The differential equation satisfied by the butchered propagator,

$$D_L\hat{S}_f(x,y) = - D_L P_f(x,y) \; ; \qquad x \neq y \tag{40}$$

implies

$$i\partial_\mu \,\mathrm{tr}\,\{\tilde{\sigma}^\mu \hat{S}_f(x,x)\} = \mathrm{tr}\,\{ D^x_L P_f(x,y) + P_f(x,y)\overleftarrow{D}^y_L \}_{x=y} . \tag{41}$$

In the two-dimensional case, the polynomial $P_f(x,y)$ was explicitly given above. Evaluating the derivatives and taking the trace over the Weyl matrices one finds ($\epsilon_{01}=1$):

$$\partial_\mu \, \mathrm{tr}\{\tilde{\sigma}^\mu \, \hat{S}_f(x,x)\} = -\frac{i}{2\pi}\,\epsilon^{\mu\nu}\partial_\mu f_\nu(x) \; . \tag{42}$$

Inserting this result in (39), we obtain

$$\ln \det D_L' = \ln \det D - \frac{i}{4\pi}\int dx \; \alpha \; \epsilon^{\mu\nu}\partial_\mu f_\nu \; . \tag{43}$$

The above calculation shows why the renormalized determinant is not gauge invariant. To specify the determinant, we had to remove a certain piece from the external field propagator. As a result of this operation, the remainder fails to satisfy the equation of motion - this in turn implies that the renormalized current $\mathrm{tr}\{\tilde{\sigma}^\mu \, \hat{S}_f(x,x)\}$ is not conserved.

The anomaly of $Z_L(f)$ contains additional terms, because the local polynomial $\int dx \; f_\mu f^\mu$ appearing in (37) is not gauge invariant:

$$Z_L(f+\partial\alpha) = Z_L(f) - \frac{1}{4\pi}\int dx \; \alpha \; \{\epsilon^{\mu\nu}\partial_\mu f_\nu - \partial_\mu f^\mu - \tfrac{1}{2}\,\Box\alpha\} \; . \tag{44}$$

It is a straightforward matter to extend this analysis to an arbitrary number of space-time dimensions. The calculation of the anomaly boils down to an analysis of the short distance singularities of the fermion propagator in an external field.

6. RIGHT-HANDED CURRENT

The generating functional of the right-handed current is given by the determinant of the right-handed Weyl operator

$$e^{iZ_R(f)} = \det D_R \; ; \quad D_R = -i\sigma^\mu(\partial_\mu - if_\mu) \; . \tag{45}$$

The operator D_R involves the matrices σ^μ instead of the matrices $\tilde{\sigma}^\mu$ which occur in D_L (if d is even, the two sets of matrices constitute inequivalent representations of the Weyl algebra).

The generating functional of the vector current is given by the determinant of the Dirac operator:

$$e^{iZ_V(f)} = \det \not{D} \; ; \quad \not{D} = -i\gamma^\mu(\partial_\mu - if_\mu) = \begin{bmatrix} 0 & D_L \\ D_R & 0 \end{bmatrix} . \tag{46}$$

It is possible to renormalize the determinants in such a manner that the product rule

$$\det \not{D} = \det D_R \cdot \det D_L \tag{47}$$

holds, i.e.

$$Z_V(f) = Z_R(f) + Z_L(f) \; . \tag{48}$$

The difference $Z_R(f) - Z_L(f)$ is the generating functional of the axial current.

As discussed above, the generating functional of the vector current can be chosen to be gauge invariant. Under a gauge transformation, the determinant of D_R must therefore pick up a phase opposite to the phase of det D_L; the anomalies in Z_R and in Z_L are of opposite sign.

As emphasized by Jackiw[5], the convention (48) is not a natural one in d=2. The natural renormalization of Z_L only involves f_+ while Z_R only contains f_-. If one uses (48) to define Z_V, one obtains a generating functional for the vector current which fails to be gauge invariant (the contributions to the anomalies of Z_R, Z_L proportional to the ϵ-tensor cancel, but the contributions proportional to $g_{\mu\nu}$ do not). The disease is easily cured by adding a local term proportional to $\int dx\, f_\mu f^\mu$; the relation (48) then, however, fails to hold. The essential point here is that the generating functionals are well defined only up to local polynomials. Some of their properties, like the form (43) of the anomaly or the product rule (48), do not hold for every choice of these local polynomials.

7. FERMIONS WITH INTERNAL QUANTUM NUMBERS

It is well known that one can bring all fermion fields to left-handed form. Instead of a single four-component Dirac spinor ψ_e describing right- and left-handed electrons and positrons, e.g., one can use two-component Weyl spinors χ_{e+}, χ_{e-}, in terms of which the Dirac field takes the form

$$\psi_e = \begin{pmatrix} \epsilon\chi^*_{e^+} \\ \chi_{e^-} \end{pmatrix} . \tag{49}$$

We collect all fermions in one left-handed field $\chi_a(x)$, where the internal quantum number $a = 1,\dots,N$ labels the various different flavours and colours. To collect all Green's functions of the corresponding free left-handed currents

$$L^\mu_{ab} = :\chi^+_a \tilde{\sigma}^\mu \chi_b: \tag{50}$$

we need an $N \times N$ matrix $f_\mu(x)_{ab}$ of external fields:

$$e^{iZ_L(f)} = \langle 0 | T e^{i\int dx \, \mathrm{tr}(f_\mu L^\mu)} | 0 \rangle . \tag{51}$$

The corresponding Weyl operator

$$D_L = -i\tilde{\sigma}^\mu(\partial_\mu - if_\mu) \tag{52}$$

is a matrix in spin space as well as in the space of internal quantum numbers. Again, the generating functional is given by the determinant of this operator

$$e^{iZ_L(f)} = \det D_L . \tag{53}$$

To discuss the Ward identities associated with the conservation of L^μ_{ab}, we consider the gauge transformation

$$f'_\mu = Uf_\mu U^+ + iU\partial_\mu U^+ \; ; \quad D'_L = UD_L U^+ . \tag{54}$$

The transformation properties of the determinant under this transformation were first calculated by Bardeen[10]. For an infinitesimal transformation

$$U = 1 + i\alpha + \dots \; ; \quad \delta f_\mu = \nabla_\mu \alpha = \partial_\mu \alpha - i[f_\mu, \alpha] \tag{55}$$

the result is

$$\delta \ln \det D_L = \frac{i}{24\pi^2} \int dx\, \epsilon^{\mu\nu\rho\sigma} \operatorname{tr}[\alpha \{\partial_\mu f_\nu \partial_\rho f_\sigma - \tfrac{1}{2} \partial_\mu (f_\nu f_\rho f_\sigma)\}], \quad (56)$$

where the trace extends over the $N \times N$ matrices α, f_μ. This result shows that, in four dimensions, there are anomalies in the Ward identities only in the case of the three- and of the four-point-functions. For N=1, the second term in (56) vanishes and the expression reduces to (14) as it should. (There are no anomalies in the four-point-function if there is only a single fermion flavour.) Note that the expression in the curly bracket is not gauge covariant. This is necessarily so: as was shown by Wess and Zumino[11], the algebraic structure of the anomalies is determined by the integrability conditions which are not consistent with a covariant expression for the anomalies.

The structure of the chiral anomalies in higher dimensions is described in Refs.[6,7,12]. The short distance approach to the problem was pioneered by Fujikawa[13]. An analysis of the d-dimensional Weyl determinant along the lines sketched in Section 5 is given in Ref. [14].

8. FERMION MASS TERMS

In the preceeding discussion we have exclusively dealt with free massless fermions. In the presence of mass terms, the free currents in general fail to be conserved. For the left-handed current connecting u and d quarks, e.g., we have

$$\partial_\mu \{\bar{u}\gamma^\mu \tfrac{1}{2}(1-\gamma_5)d\} = \tfrac{i}{2}\, \bar{u}\, \{(m_u - m_d) + \gamma_5 (m_u + m_d)\}\, d\ . \quad (57)$$

Even if there were no anomalies, the Green's functions of this current therefore fail to be conserved. To keep trace of the mass terms in the generating functional, one introduces external scalar and pseudoscalar fields and investigates the transformation properties of the generating functional in the presence of these additional fields[15]. One finds that the leading short distance singularities of the fermion propagator, which are responsible for the occurrence of anomalies, are not affected by scalar or pseudoscalar fields. Although mass terms do generate a right-hand side in the Ward identities for the various currents, they do not modify the anomalies, which can be analyzed on the basis of the corresponding massless theory. In the following, I shall therefore drop all mass terms and disregard the Higgs field which generates them.

9. FROM EXTERNAL FIELDS TO INTERACTING GAUGE FIELDS

In the standard model, the fermions interact with a set of gauge fields $G^i_\mu(x)$ belonging to the gauge group SU(3)xSU(2)xU(1). Let us denote the generators of the gauge group on the left-handed fermion fields by T_i, such that the gauge fields interact with the fermions through the matrix

$$G_\mu(x) = \sum_i G^i_\mu(x)\ T_i\ . \tag{58}$$

The Lagrangian of the standard model is of the form

$$\mathcal{L} = \mathcal{L}_G + i\chi^+\ \tilde{\sigma}^\mu(\partial_\mu - iG_\mu)\ \chi\ +\ \mathcal{L}_H\ . \tag{59}$$

The quantity $\mathcal{L}_G$ stands for the gauge field Lagrangian $\sim g^{-2}\ \mathrm{tr}\ G_{\mu\nu}G^{\mu\nu}$. The term $\mathcal{L}_H$ represents the part of the Lagrangian which involves the Higgs field. For the reason given in the last section, I disregard this part and consider the path integral

$$\int [dG][d\chi]e^{i\int dx\mathcal{L}_G - i\int dx\chi^+ D_L\chi}\ ;\quad D_L = -i\tilde{\sigma}^\mu(\partial_\mu - iG_\mu)\ . \tag{60}$$

Integrating the fermions out, this becomes

$$\int [dG]e^{i\int dx\mathcal{L}_G}\ \det D_L\ . \tag{61}$$

For the consistency of perturbation theory (graphs involving gauge fields as internal lines) it is crucial that gauge invariance can be maintained order by order. This is the case only if $\det D_L$ is invariant under the transformations generated by the gauge group G, i.e., if there are no anomalies in the Ward identities for the currents $\chi^+\ \tilde{\sigma}^\mu T_i\ \chi$.

A simple model for which there is an anomaly that ruins gauge invariance is Quantum Electrodynamics with only left-handed electrons (and right-handed positrons):

$$\mathcal{L} = -\frac{1}{4e^2}\ G_{\mu\nu}G^{\mu\nu} + i\chi^+\ \tilde{\sigma}^\mu(\partial_\mu - iG_\mu)\ \chi\ . \tag{62}$$

For ordinary QED there is no problem: the anomalies generated by the left-handed electron field cancel the anomalies produced by the left-handed positron field. Gauge fields coupled to fermions through vector currents are always anomaly free.

It is not difficult to find out under what conditions cancellation of anomalies takes place. A glance at (56) shows that infinitesimal gauge transformations of the form $\alpha=\Sigma_i\alpha^i T_i$ are anomaly free if and only if the fermion representation satisfies the condition

$$\mathrm{tr}(T_i\{T_k,T_l\}) = 0 \ . \tag{63}$$

In QED there is a single generator, $T_i = Q$, where Q is the electric charge matrix and the condition (63) amounts to $\mathrm{tr}\ Q^3 = 0$. For left-handed QED, Q is a 1×1 matrix with $\mathrm{tr}\ Q^3 \neq 0$ and the condition is violated, whereas for the standard electromagnetic interaction of the electron there are two left-handed fields with opposite charge such that $\mathrm{tr}\ Q^3 = 0$.

In the standard model based on the gauge group SU(3)×SU(2)×U(1) the condition (63) would be violated if there were only quarks, but no leptons (or vice versa). The condition is obeyed if the fermions occur in copies of the first generation (ν_e,e,u,d), provided the electric charge of the quarks is related to the electric charge of the electron in the familiar manner. In order for the standard model to be a consistent gauge theory, it is thus necessary that the hydrogen atom is electrically neutral, $Q_e + 2Q_u + Q_d = 0$.

10. ANOMALIES IN EXTERNAL CURRENTS

Suppose that those currents to which the gauge fields are coupled are anomaly free. What about the remaining currents which do not occur in the Lagrangian of the theory? To analyze the properties of these 'external' currents we again introduce a set of external fields and consider the generating functional

$$e^{iZ_L(f)} = \int [dG] e^{i\int \mathcal{L}_G dx} \det D_L \ ; \quad D_L = - i\tilde{\sigma}^{\mu}(\partial_{\mu}-iG_{\mu}-if_{\mu}) \ . \tag{64}$$

In order for the gauge invariance associated with the gauge group G to remain intact, we have to restrict ourselves to gauge invariant external currents, i.e.

$$[T_i, f_\mu] = 0 \ . \tag{65}$$

The Ward identities obeyed by the external currents are easily worked out from the general formula (56) with the replacement $f_\mu \to G_\mu + f_\mu$. They involve terms proportional to the winding number density $\epsilon^{\mu\beta\rho\sigma} G_{\mu\nu} G_{\rho\sigma}$ of the gauge fields.

Why are the Ward identities for external currents of more than academic interest? The point is that conserved currents correspond to symmetries of the Hamiltonian. If the conservation of an external current is found to be ruined by an anomaly, the corresponding symmetry is lost. For the standard model, anomalies in external axial currents are vital for our understanding of the decay $\pi^0 \to 2\gamma$ and of the mass spectrum of the low lying pseudoscalar mesons.

11. GRAVITATIONAL ANOMALIES

Finally, I turn to fermions moving in an external gravitational field. In a curved space, the Dirac algebra

$$\{\gamma^\mu, \gamma^\nu\} = 2g^{\mu\nu}(x) \tag{66}$$

leads to coordinate dependent γ-matrices which can be represented in the form

$$\gamma^\mu(x) = e^\mu_a(x)\gamma^a \ , \tag{67}$$

where γ^a is a representation of the Dirac algebra in flat space and where the vielbein e^μ_a obeys

$$g^{\mu\nu}(x) = e^\mu_a(x) e^\nu_b(x) \overset{o}{g}{}^{ab} \ . \tag{68}$$

The vielbein is determined by the geometry $g_{\mu\nu}(x)$ only up to a local Lorentz transformation ('frame rotation')

$$e^\mu_a(x)' = \Lambda^{-1}(x)^b_a \, e^\mu_b(x) \ . \tag{69}$$

The corresponding change in the γ-matrices is given by

$$\gamma^{\mu}(x)' = S[\Lambda(x)]\gamma^{\mu}(x)S[\Lambda(x)]^{-1} , \tag{70}$$

where $S[\Lambda]$ is the spinor representation associated with the Lorentz transformation Λ.

For massless fermions in an external gravitational field, the Dirac equation reads

$$-i\gamma^{\mu}(x) \{\partial_{\mu}+\omega_{\mu}(x)\} \psi(x) = 0 . \tag{71}$$

The spin connection ω_{μ} insures invariance under local frame rotations:

$$\omega_{\mu}' = S\omega_{\mu}S^{-1} + S\partial_{\mu}S^{-1} . \tag{72}$$

The connection can be expressed in terms of the vielbein and its first derivative as

$$\omega_{\mu} = \tfrac{1}{4} \gamma^{\nu} \{\partial_{\mu}\gamma_{\nu}-\Gamma^{\lambda}_{\mu\nu}\gamma_{\lambda}\} , \tag{73}$$

where $\Gamma^{\lambda}_{\mu\nu}$ is the Christoffel symbol associated with the metric $g_{\mu\nu}$.

Suppose that the space is asymptotically flat: $e^{\mu}_{a} \to \delta^{\mu}_{a}$ as $x^0 \to \pm \infty$. What is the probability for the gravitational field not to create any fermion pairs? The probability amplitude is given by

$$\langle 0 \text{ out} | 0 \text{ in}\rangle = \det \not{D} ; \quad \not{D} = -i\gamma^{\mu}(\partial_{\mu}+\omega_{\mu}) . \tag{74}$$

Assume that the geometry $g_{\mu\nu}(x)$ is given. To calculate the transition amplitude, the geometry alone is, however, not sufficient: the Dirac operator explicitly contains the vielbein field e^{a}_{μ}. We therefore need to know whether any vielbein which describes the same geometry also leads to the same transition amplitude. Vielbeins belonging to the same geometry differ at most by a frame rotation. Under a frame rotation, the Dirac operator transforms according to

$$\not{D}' = S \not{D} S^{-1} . \tag{75}$$

The problem therefore boils down to the question of whether or not the determinant of $\not{D}$ is invariant under the transformation (75). For the Dirac operator, this is indeed the case: the transition amplitude only depends on the geometry. For left-handed fermions, however, there is a problem[4,7,16]: the determinant of the Weyl operator is not invariant under frame rotations if the dimension d of space-time is of the form $d = 4n+2 = 2,6,10,\ldots$.

To understand the origin of this 'Lorentz anomaly', it is instructive to again consider a space-time of two dimensions[17,18]. In d=2, the σ-matrices are numbers. In flat space, we have $\tilde{\sigma}^0 = \tilde{\sigma}^1 = 1$ and the free Weyl operator reduces to $D_L = -i(\partial_0+\partial_1)$. In curved space it is of the form

$$D_L = -i\tilde{\sigma}^{\mu}(x)\ \{\partial_{\mu}+\omega_{\mu}(x)\}\ . \tag{76}$$

The vector $\tilde{\sigma}^{\mu}(x)$ is a linear combination of the zweibein vectors e^{μ}_{0} and e^{μ}_{1} :

$$\tilde{\sigma}^{\mu}(x) = e^{\mu}_{0}(x) + e^{\mu}_{1}(x)\ ; \quad \sigma^{\mu}(x) = e^{\mu}_{0}(x) - e^{\mu}_{1}(x)\ . \tag{77}$$

The spin connection is given by a derivative of the zweibein vectors

$$\omega_{\mu} = \tfrac{1}{4}\, g_{\mu\nu}(\tilde{\sigma}^{\alpha}\partial_{\alpha}\sigma^{\nu}-\sigma^{\alpha}\partial_{\alpha}\tilde{\sigma}^{\nu})\ . \tag{78}$$

Under a frame rotation

$$\tilde{\sigma}^{\mu\prime} = e^{2\lambda}\,\tilde{\sigma}^{\mu}\ ; \qquad \omega_{\mu}{}' = \omega_{\mu} + \partial_{\mu}\lambda\ . \tag{79}$$

The spin connection thus plays a role analogous to an imaginary electromagnetic field; a frame rotation changes the spin connection by a gauge transformation.

In two dimensions, one can always bring the metric to conformally flat form, $g_{\mu\nu}(x) = F(x)\ \overset{o}{g}_{\mu\nu}$. This property allows one to explicitly calculate the determinant of the Weyl operator[19]. In the presence of both an external gravitational field and an external electromagnetic field $G_{\mu}(x)$,

$$D_L = -i\tilde{\sigma}^{\mu}\ \{\ \partial_{\mu}+\omega_{\mu}-iG_{\mu}\ \} \tag{80}$$

the vacuum-to-vacuum transition amplitude is given by[18]

$$\det D_L = \exp\ \{\tfrac{i}{3}\, Z_g(\omega)+iZ_g(G)\}\ , \tag{81}$$

where the functional $Z_g(f)$ is quadratic in the vector field f_μ :

$$Z_g(f) = \frac{1}{8\pi} \int dx\, dy\, \hat{f}(x) \Delta_g(x,y) \hat{f}(y) \; ; \qquad \hat{f}(x) = \partial_\mu \{ (\epsilon^{\mu\nu} - \sqrt{-g}\, g^{\mu\nu}) f_\nu(x) \} \; . \quad (82)$$

The kernel $\Delta_g(x,y)$ denotes the scalar Feynman propagator in an external gravitational field

$$\partial_\mu \sqrt{-g}\, g^{\mu\nu} \partial_\nu \Delta_g(x,y) = \delta(x-y) \; . \qquad (83)$$

Note that $\ln \det D_L$ is unique only up to a local polynomial in the external fields. The expression given in Ref.[18] differs from (82) by a local term proportional to $\int dx \sqrt{-g}\, f^\mu_\mu f^\mu$ (compare Section 5).

In flat space, the spin connection vanishes and the term involving $Z_g(\omega)$ is therefore absent. Furthermore, we have

$$\partial_\mu \{ (\epsilon^{\mu\nu} - \sqrt{-g}\, g^{\mu\nu}) f_\nu \} = -\partial_- f_+ \qquad (84)$$

such that $Z_g(G)$ reduces to the generating functional (20) associated with the left-handed current in flat space, as it should.

The determinant of the Weyl operator is invariant neither under gauge transformations of the electromagnetic field nor under local Lorentz transformations, because the functional $Z_g(f)$ is not invariant under $f_\mu \to f_\mu + \partial_\mu \alpha$. The transition amplitude therefore not only fails to be independent of the gauge chosen for the electromagnetic field, it also depends on the zweibein chosen to represent the geometry. The failure of $\det D_L$ to be invariant under frame rotations is referred to as a Lorentz anomaly. The problem only shows up in the phase of the transition amplitude - the transition probability only depends on the geometry and is invariant under gauge transformations of the electromagnetic field.

The Lorentz anomaly also shows up in the properties of the energy-momentum tensor, which determines the response of the system to a change in the metric

$$\delta \ln \langle 0 \text{ out} | 0 \text{ in} \rangle = -\tfrac{i}{2} \int dx\, \delta g_{\mu\nu}(x)\, T^{\mu\nu}(x) . \tag{85}$$

Since the transition amplitude is not fixed by the geometry alone, but explicitly depends on the vielbein, this formula does not specify $T^{\mu\nu}(x)$ unambiguously. What is well defined is the response of the system to a deformation of the vielbein

$$\delta \ln \langle 0 \text{ out} | 0 \text{ in} \rangle = i \int dx\, \delta e^{\mu}_{a}(x)\, t^{a}_{\mu}(x) . \tag{86}$$

The deformation of the metric fixes δe^{μ}_{a} up to an infinitesimal frame rotation. Since an infinitesimal frame rotation at the point x produces a change in the transition amplitude which only depends on the properties of the vielbein in the vicinity of the point x, the quantity $T^{\mu\nu}$ is well defined up to a symmetric local polynomial $p^{\mu\nu} = p^{\nu\mu}$,

$$T^{\mu\nu} = t^{\underline{\mu\nu}} + p^{\mu\nu} , \tag{87}$$

where $t^{\underline{\mu\nu}}$ is the symmetric part of the tensor $t^{\mu\nu} = g^{\mu\alpha} e^{\nu}_{a} t^{a}_{\alpha}$.

If there is an external electromagnetic field, there is no reason for the energy-momentum tensor to be conserved, because energy may flow in and out of the external electromagnetic field. Let us, therefore, switch the electromagnetic field off and consider pure gravity. If the gravitational field $g_{\mu\nu}(x)$ is the only external field, coordinate invariance implies that $T^{\mu\nu}$ is covariantly conserved

$$\nabla_{\mu} T^{\mu\nu} = 0 . \tag{88}$$

It turns out, however, that independently of how one chooses the local polynomial $p^{\mu\nu}$, this relation fails to be satisfied. In two dimensions, one instead finds

$$\nabla_{\mu} T^{\mu\nu} = \frac{1}{96\pi} \epsilon^{\mu\nu} \partial_{\nu} R , \tag{89}$$

where R is the scalar curvature.

At first sight, this relation appears to indicate a breakdown of coordinate invariance. Since the quantity $Z_g(\omega)$ is, however, explicitly coordinate invariant, this is not the proper conclusion to draw. Instead, the relation (89) expresses the fact that energy flows in and out of the degree of freedom which specifies the direction of the zweibein - the metric is not the only external field the system interacts with.

13. LORENTZ ANOMALIES VERSUS COORDINATE ANOMALIES

As was shown by Bardeen and Zumino[16], one can always modify the determinant of the Weyl operator in such a fashion that it does become frame independent - at the cost of coordinate invariance. In fact, in d=2, it suffices to add the local functional[17]

$$F = \frac{i}{24\pi} \int dx\sqrt{-g}\ \{\tfrac{1}{4} R \ln[k_\mu \tilde{\sigma}^\mu(x)] + g^{\mu\nu}\omega_\mu\omega_\nu\} \qquad (90)$$

to $\ln \det D_L$. The quantity

$$\ln \overline{\det D_L} = \ln \det D_L + F \qquad (91)$$

is invariant under local Lorentz transformations, but breaks coordinate invariance through the constant vector k_μ . In other words, $\overline{\det D_L}$ does not have a Lorentz anomaly; instead it has a coordinate anomaly.

The functional F is local, but it is not a polynomial. The short distance singularities of the Weyl operator only generate polynomial ambiguities in $\ln \det D_L$ (local polynomials of the vielbein matrix, of its inverse and of its derivatives.) The counter terms needed are polynomials - the modification given above is therefore outside the class of counter terms which occur in the renormalization procedure. The short distance singularities break frame independence, but they do not break coordinate independence. In this sense, Lorentz anomalies and coordinate anomalies are not equivalent. (These statements also hold in higher dimensions[20]).

Left-handed fermions generate a Lorentz anomaly in $d = 2,6,10,\ldots$ The phenomenon does not occur in four dimensions, but it does occur in

Kaluza-Klein theories. The Lorentz anomaly is a purely gravitational effect; the manner in which the phase of the vacuum-to-vacuum transition amplitude changes under a frame rotation only depends on the curvature of the space and is not modified if gauge fields are present. On the other hand, the gravitational field does modify the structure of the chiral anomalies discussed in the first part of this lecture (there are no mixed Lorentz anomalies, but there are mixed chiral anomalies). Chiral anomalies occur whenever the dimension of space-time is even. In four dimensions, e.g., the Ward identity for the left-handed U(1) current $\chi^{+}\tilde{\sigma}^{\mu}\chi$ recieves a purely gravitational contribution proportional to the square of the Riemann tensor[21]. An external gravitational field therefore ruins gauge invariance, unless the generators of the fermion representation are traceless, $\mathrm{tr}\ T_i = 0$. If this condition is not satisfied, the quantum theory of the gauge field does not survive the perturbation produced by an external gravitational field (in the standard model the condition holds, generation by generation).

In the case of the gravitational field, a consistent perturbative quantization is not available, even if there are no fermions. It is therefore not possible to extend the preceeding discussion from external gravitatio?‘l fieds to interacting gravitational fields. Superstrings may provide a consistent framework within which gravity as we know it emerges as an effective low energy approximation. At low energies, the internal degrees of freedom of the string are frozen and the theory reduces to a local field theory involving a rich spectrum of fields - a gravitational field, as well as gauge fields and fermions. For some of these theories, the structure of the string Lagrangian implies that the anomalies in the corresponding effective low energy theory cancel[22] - as if a consistent theory for quantum gravity was not enough of a miracle already.

ACKNOWLEDGEMENTS

It is a pleasure to thank Joachim Debrus and Allen Hirshfeld for their kind hospitality at Bad Honnef. Furthermore, I am indebted to Ph. Martin for the permission to reprint this article from Helvetica Physica Acta 59: 201 (1986).

REFERENCES

1. S. L. Adler, Phys. Rev. 117: 2426 (1969);

   J. S. Bell and R. Jackiw, Nuovo Cim. 60A: 47 (1969).

2. D. J. Gross and R. Jackiw, Phys. Rev. D6: 447 (1972);

   C. Bouchiat, J. Iliopoulos and Ph. Meyer, Phys. Lett. 38B: 519 (1972);

   C. P. Korthals Altes and M. Perrottet, Phys. Lett. 39B: 546 (1972);

   H. Georgi and S. Glashow, Phys. Rev. D6: 429 (1972);

   H. Georgi, Nucl. Phys. B156: 126 (1979);

   G. `t Hooft in "Recent Developments in Gauge Theories",

   G.`t Hooft et al., eds., Plenum Press, New York (1980);

   A. Zee, Phys. Lett 99B: 110 (1981).

3. E. Witten, Phys. Lett. 117B: 324 (1982).

4. L. Alvarez-Gaumé and E. Witten, Nucl. Phys. B234: 269 (1984).

5. M.F.Atiyah and I.M. Singer, Proc. Nat. Acad. Sci. USA 81: 2597 (1984);

   R. Stora, Cargése Lectures 1983, Plenum Press, New York, to be publ.;
   L. Baulieu, ibid and Nucl. Phys. B241: 557 (1984);

   B. Zumino, Les Houches Lectures 1983, R. Stora and B. De Witt, eds., North-Holland, to be published; R. Jackiw, ibid;

   O. Alvarez, I. M. Singer and B. Zumino, Comm. Math. Phys. 96: 409 (1984);

   L. Alvarez-Gaumé and P. Ginsparg, Nucl. Phys. B243: 449 (1984).

6. B. Zumino, Wu Yong-Shi and A. Zee, Nucl. Phys. B239: 477 (1984).

7. L. Alvarez-Gaumé and D. Ginsparg, Harvard preprint HUTP 84/A016.

8. J. Schwinger, Phys. Rev. 128: 2425 (1962);

   K. Johnson, Phys. Lett. 5: 253 (1963);

   for a review, see R. Jackiw, ref [5].

9. J. Schwinger, Phys. Rev. 82: 664 (1951).

10. W. A. Bardeen, Phys. Rev. 184: 1848 (1969).

11. J. Wess and B. Zumino, Phys. Lett. 37B: 95 (1971).

12. P.H. Frampton and T.W. Kephart, Phys. Rev. Lett. 50: 1343,1347 (1983).

13. K. Fujikawa, Phys. Rev. Lett. 42: 1195 (1979); 44: 1733 (1980); Phys. Rev. D21: 2848 (1980); D22: 1499(E) (1980); D23: 2262 (1981); D29: 285 (1984);
A. P. Balachandran, G. Marmo, V.P. Nair and C.G. Trahern, Phys. Rev. D25: 2713 (1982);
T. Matsuki, Phys. Rev. D28: 2107 (1983).
14. H. Banerjee and R. R. Banerjee, preprints Saha Institute of Nuclear Physics, SINP-TNP-84/6 and 85/2, Calcutta;
H. Leutwyler, Phys. Lett. 152B: 78 (1985) and preprint University of Bern BUTP84/33, to be published in "Quantum Field Theory and Quantum Statistics, essays in honour of E.S. Fradkin" (Adam Hilger Publ. Co.).
15. J. Gasser and H. Leutwyler, Ann of Phys 158: 142, Appendix A (1984).
16. W.A. Bardeen and b Zumino, Nucl. Phys. B244: 421 (1984);
L. Baulieu and J. Thierry-Mieg, Phys. Lett. 145B: 53 (1984);
F. Langouche, T. Schücker and R. Stora, Phys. Lett. 145B: 342 (1984).
17. F. Langouche, Phys. Lett. 148B: 93 (1984).
18. H. Leutwyler, Phys. Lett. 153B: 65 (1985); 155B: 469(E) (1985).
19. A. M. Polyakov, Phys. Lett. 103B: 207,211 (1981);
E. S. Fradkin and A. A. Tseytlin, Phys. Lett. 106B: 63 (1981).
20. H. Leutwyler and S. Mallik, preprint University of Bern BUTP 85/23, to be published.
21. R. Delbourgo and A. Salam, Phys. Lett. 40B: 381 (1972);
T. Eguchi and P. G. O. Freund, Phys. Rev. Lett. 37: 1251 (1976).
22. M. B. Green and J. H. Schwarz, Phys. Lett. 149B: 117 (1984).

GAUGE INVARIANCE IN SPITE OF ANOMALIES

N. K. Falck*

II Institut für Theoretische Physik
der Universität Hamburg
Notkestr. 85, 2000 Hamburg 52, Fed. Rep. Germany

ABSTRACT

We show that, contrary to previous belief, even in models with chiral fermions coupled to a dynamical gauge field an improved quantization procedure leads to a quantum system which is gauge (BRST) invariant; there are no genuine anomalies which spoil this invariance. This statement is proven in the path integral formalism for the general case, and is made explicit for the chiral Schwinger model, which leads to a consistent quantum theory in spite of the "anomaly".

* Supported by the Bundesministerium für Forschung und Technologie, 05 4 HH 92 P/3, Bonn, FRG

1. INTRODUCTION

Prof. Leutwyler indicated in his lecture at this School[1] that anomalies lead to difficulties in the quantization of a gauge field which couples to chiral fermions. In the present talk I wish to show how these difficulties might be resolved.

Let us start with a brief review of the reason why such a theory seems to be inconsistent. The anomaly is defined as the (covariant) divergence of a current:

$$\alpha = D_\mu J^\mu . \tag{1.1}$$

In the following I shall always consider that current which couples to the gauge field, since only the divergence of this one is dangerous. We are only interested in chiral theories, hence we take our Lagrangian to be:

$$L = -\tfrac{1}{4} F^a_{\mu\nu} F^{\mu\nu a} + \bar\psi[i\partial\!\!\!/ - eA\!\!\!/\tfrac{1}{2}(1+i\gamma_5)]\psi . \tag{1.2}$$

In this case the current is given by:

$$J_\mu^{\ a} = \bar\psi\gamma^\mu\tau^a\tfrac{1}{2}(1+i\gamma_5)\psi , \tag{1.3}$$

where τ^a are the generators of the gauge group, normalized according to $\mathrm{tr}\ \tau^\alpha\tau^\beta = \frac{1}{2}\ \delta^{\alpha\beta}$. The matrix valued field A in Eq. (1.2) is defined as $A_\mu = A_\mu^{\ a}\tau^a$ and Eq. (1.1) has to be understood as a matrix equation in the nonabelian case. The equations of motion with respect to the gauge field read

$$D_\mu F^{\mu\nu} = J^\nu ., \tag{1.4}$$

which are the nonabelian generalization of the inhomogeneous Maxwell equations. Applying D_ν to both sides of eq. (1.4) yields:

$$D_\nu D_\mu\ F^{\mu\nu} = D_\nu J^\nu . \tag{1.5}$$

The problem now is that the left-hand side $D_\nu D_\mu F^{\mu\nu}$ vanishes identically (it is proportional to $[F_{\mu\nu}, F^{\mu\nu}]$). Hence Eq. (1.5) is consistent only if $D_\nu J^\nu = \alpha = 0$, i.e. if there is no anomaly. However, as Prof. Leutwyler has pointed out, $\alpha \neq 0$ seems to be unavoidable in most chiral models with quantized fermions.

This observation led to the general conviction that anomalous gauge theories are inconsistent and should be avoided. Historically, this conviction arose in the following way. The first time anomalies were discussed in the literature was in connection with an inconsistency in pertubation theory, where the Ward identities recieved anomalous contributions due to the impossibility of maintaining gauge invariance order by order[2]. Anomalous Ward identities mean that the renormalization program does not go through for these theories. However, it does not seem to make much sense to speak about renormalizability if the theory has mortal defects even before the quantization procedure for the gauge field has been carried through. This means that the inconsistency in the inhomogeneous Maxwell equations has to be considered to be the more fundamental one.

The requirement of anomaly freedom emerged to be a guiding principle for building physical models. If the gauge group is not safe by itself, the particle content in the fermion sector is chosen in such a way that all anomalies cancel. This principle is construed as the most compelling theoretical argument for the existence of the top quark and for the relevance of superstring models.

This describes the situation up to about 1985. In that year Jackiw and Rajaraman[3] showed that anomalies do not necessarily spoil consistency. The chiral Schwinger model turned out to be a counterexample; it leads to a consistent and unitary quantum theory in spite of the apparent presence of an anomaly. If this behaviour should be a general feature we are forced to change our attitude concerning anomalous gauge theories completely. Indeed, in 1986 some interesting mechanisms have been proposed as to how anomalous gauge theories could be quantized consistently[4-9,34]. The most convincing one[7-9] (this is my personal point of view) consists of a careful reanalysis of the Faddeev-Popov procedure[10], which in its time was the decisive breakthrough for the quantization of gauge fields in the first place. The result of this investigation is the fact that certain scalar fields with a Wess-Zumino action[11] have been missed in earlier treatments. These scalar fields cancel the anomalies of the fermionic sector and render the theory gauge invariant even at the quantum level (in a quantized theory gauge invariance means BRST[12,13] invariance). Hence there is no inconsistency in Eq. (1.5) since the anomaly a vanishes automatically. This observation opens up for the first time the possibility of formulating a quantized chiral gauge theory without an ill-defined starting point. This quantum theory should be expected to be in a much better shape than its traditional predecessors; its consistency, unitarity and renormalizability, however, remain to be proven. Hence all calculations which traditionally led to inconsistencies should be redone in the correct framework.

This is a very difficult task, however, so not much progress has been made to date. Only the two-dimensional abelian case has been investigated in detail[3,9,14-22]. The reason for this lies in the fact that in two dimensions the fermions can be integrated out explicitly, leaving behind a purely bosonic theory which, in the abelian case, can even be solved. Hence the chiral Schwinger model (chiral QED in two dimensions) can serve as a toy model in order to gain some experience as to how the anomaly cancellation mechanism works.

The aim of this talk is two-fold: I wish to present the correct path integral approach to anomalous gauge theories and I wish to show that this formalism works succesfully in the only model which has been studied up to now: the chiral Schwinger model.

2. PATH INTEGRAL QUANTIZATION OF GAUGE THEORIES (REVISITED)

This section is devoted to the proof that a careful path integral quantization of gauge theories with chiral coupling to fermions automatically leads to an anomaly-free theory. The most important observation is that the integration should be performed over the complete configuration space of the gauge fields, including those which are related by gauge transformations[7-9]. This leads to the generating functional (more precisely, to the vacuum-to-vacuum amplitude)

$$Z = \int dA \, d\psi \, d\bar{\psi} \, e^{iS[\bar{\psi},\psi,A]} , \tag{2.1}$$

where $S[\bar{\psi},\psi,A]$ is the classical action of Eq. (1.2). Since everything that follows in this procedure is a direct consequence of Eq. (2.1) this kind of integration should be motivated. I shall come back to this point as soon as we have developed the formalism as far as is necessary.

The classical action is invariant with respect to local gauge transformations of the form ($h = e^{i\alpha}$, $\alpha = \alpha^a \tau^a$):

$$A_\mu \to A_\mu^h = h A_\mu h^{-1} + \frac{i}{e} h \partial_\mu h^{-1} ,$$

$$\psi \to \psi^h = e^{\frac{i}{2}(1+i\gamma_5)\alpha} \psi ; \quad \bar{\psi} \to \bar{\psi}^h = \bar{\psi} e^{-\frac{i}{2}(1-i\gamma_5)\alpha} . \tag{2.2}$$

In order to see the invariance property one can use the identities:

$$\partial_\mu e^{\frac{i}{2}(1+i\gamma_5)\alpha} = \tfrac{1}{2}(1+i\gamma_5)\partial_\mu h ; \quad \tfrac{1}{2}(1+i\gamma_5) e^{\frac{i}{2}(1+i\gamma_5)\alpha} = \tfrac{1}{2}(1+i\gamma_5) h \tag{2.3}$$

which are valid since $\frac{1}{2}(1+i\gamma_5)$ is a projection operator.

There is a problem with the generating functional of Eq. (2.1), namely that in most cases it is, in its original form, not useful at all. For instance it does not allow us to define a propagator for the gauge fields, since the operator between A^2 is not invertible, so it is impossible to apply perturbation theory. At this point the usual procedure is to perform the so-called Faddeev-Popov trick[10], so we have to repeat this here. To this aim an appropriate expression for unity is inserted in Eq. (2.1);

$$1 = \Delta_f[A] \int dg\ \delta(f(A^g)) . \tag{2.4}$$

Here f is a gauge fixing function and A^g is the gauge field transformed by g. The functional integration has to be performed over the gauge group and $\Delta_f[A]$ is the Faddeev-Popov determinant, defined in such a way that Eq.(2.4) holds. In the resulting expression for Z the integration variables are relabelled according to $(A,\psi,\bar{\psi}) \to (A^{g^{-1}},\psi^{g^{-1}},\bar{\psi}^{g^{-1}})$; this yields :

$$Z = \int dg\ dA^{g^{-1}} d\psi^{g^{-1}} d\bar{\psi}^{g^{-1}}\ \Delta_f[A^{g^{-1}}]\ \delta(f(A))\ e^{iS[\bar{\psi}^{g^{-1}},\psi^{g^{-1}},A^{g^{-1}}]} . \tag{2.5}$$

Now we can use the fact that dA, $\Delta f[A]$ and $S[\bar{\psi},\psi,A]$ are gauge invariant, this leads to:

$$Z = \int \mathcal{D}A\ dg\ d\psi^{g^{-1}}\ d\bar{\psi}^{g^{-1}}\ e^{iS[\bar{\psi},\psi,A]} . \tag{2.6}$$

Here

$$\mathcal{D}A = dA\ \Delta_f[A]\ \delta(f(A)) \tag{2.7}$$

is the integration measure of the gauge surface determined by the gauge fixing function f and $\Delta_f[A]$ serves as the square root of the determinant of the metric, which is necessary to form an invariant volume element.

The essential point now (and here we leave the path of the Faddeev-Popov procedure) is that the fermionic measure is *not* invariant[23], instead it generates a nontrivial Jacobian:

$$d\psi^{g^{-1}}\, d\bar{\psi}^{g^{-1}} = d\psi\, d\bar{\psi}\, e^{i\alpha_1[A,g^{-1}]} , \qquad (2.8)$$

where the 1-cocycle $\alpha_1[A,g^{-1}]$ is called the Wess-Zumino action. In order to understand the nontriviality of this transformation behaviour look at Eq. (2.2). The γ_5-independent part of the transformation of $d\bar{\psi}$ cancels that of $d\psi$, the same, however, does not happen for the γ_5 part. Then the Jacobian is essentially the functional determinant of $\exp(\gamma_5\alpha)$. This, however, is not well-defined, because it tends to be (i) infinity, due to the fact that it is a functional determinant, (ii) zero, since γ_5 is traceless. Consequently, regularization is required, that is how the gauge field A enters into α_1. Thus we end up with the final result for the generating functional:

$$Z = \int \mathcal{D}A\, dg\, d\psi\, d\bar{\psi}\, e^{iS_{st}[\bar{\psi},\psi,A,g]} , \qquad (2.9)$$

with the "standard action"[9] :

$$S_{st}[\bar{\psi},\psi,A,g] = S[\bar{\psi},\psi,A] + \alpha_1[A,g^{-1}] . \qquad (2.10)$$

For later use we rewrite the generating functional according to

$$Z = \int \mathcal{D}A\, dg\, d\psi\, d\bar{\psi}\, e^{iS[\bar{\psi},\psi,A^{g^{-1}}]} , \qquad (2.11)$$

which is identical to Eq.(2.5) if the relabelling of the fermions is not performed there.

At this point I would like to make two remarks:

(i) The variation of the Wess-Zumino action gives the anomaly: let $h(x) = 1 + \epsilon^{\alpha}(x)\,\tau^{\alpha}$ be an infinitesimal gauge transformation, then

$$\alpha_1[A^h, h^{-1}g^{-1}] - \alpha_1[A,g^{-1}] = -\int dx\, \epsilon^a(x)\, a^a(x) . \qquad (2.12)$$

(ii) In the original Faddeev-Popov procedure the fermionic measure has been considered to be gauge invariant, which is correct for the case of vector-like gauge boson-fermion coupling, where the gauge transformation of the fermions does not involve γ_5. Then the integrand of Eq. (2.9) is independent of g, hence the group integration only gives an (infinite) irrelevant constant, which can be absorbed in the normalization of Z. In this case the generating functional reads

$$Z = \int \mathcal{D}A\, d\psi\, d\bar{\psi}\, e^{iS[\bar{\psi},\psi,A]} , \qquad (2.13)$$

which is the classical result of Faddeev and Popov [10]. If the theory has an anomaly, α_1 does depend on g and the g-integration becomes nontrivial. In this case Eq. (2.13) is incorrect and should not be applied.

Now we have to discuss the anomaly. To this aim we define the effective actions:

$$e^{i\tilde{W}[A]} = \int d\psi\, d\bar{\psi}\, e^{iS_{st}[\bar{\psi},\psi,A,g]} , \tag{2.14}$$

$$e^{iW[A]} = \int d\psi\, d\bar{\psi}\, dg\, e^{iS_{st}[\bar{\psi},\psi,A,g]} . \tag{2.15}$$

According to the generating functional of Eq.(2.9) the anomaly has to be calculated from W[A] rather than from $\tilde{W}$[A], since the g field contributes to the current, too. Hence, symbolically,

$$\alpha = \frac{\delta W[A]}{\delta \epsilon} , \tag{2.16}$$

where ϵ^α are the gauge parameters. Now we shall see that W(A) is gauge invariant, i.e. that the anomaly vanishes. We start with the gauge transformation of $\tilde{W}$[A]:

$$e^{i\tilde{W}[A^h]} = \int d\psi\, d\bar{\psi}\, e^{iS[\bar{\psi},\psi,A^h]} . \tag{2.17}$$

Relabelling the integration variables $(\psi,\bar{\psi}) \rightarrow (\psi^h,\bar{\psi}^h)$, using Eq. (2.8), and exploiting the fact that S is gauge invariant, we find:

$$\alpha_1[A,h] = \tilde{W}[A^h] - \tilde{W}[A] , \tag{2.18}$$

which can be used to prove the so-called "one cocycle condition" for α_1:

$$\alpha_1[A^h,g^{-1}] + \alpha_1[A,h] = \alpha_1\,[A,hg^{-1}] . \tag{2.19}$$

Hence we can calculate

$$e^{iW[A^h]} = \int d\psi\, d\bar{\psi}\, dg\, e^{i\{S[\bar{\psi},\psi,A^h] + \alpha_1[A^h,g]\}} . \tag{2.20}$$

The same steps as those leading to Eq. (2.18) yield

$$e^{iW[A^h]} = \int d\psi\, d\bar{\psi}\, dg\, e^{i\alpha_1[A,h]}\, e^{i\{S[\bar{\psi},\psi,A] + \alpha_1[A^h,g^{-1}]\}}$$

$$= \int d\psi\, d\bar{\psi}\, d(gh^{-1})\, e^{i\{S[\bar{\psi},\psi,A] + \alpha_1[A,hg^{-1}]\}} = e^{iW[A]} . \tag{2.21}$$

Here we have used Eq. (2.19) and the invariance of the group measure. Thus we may conclude that the effective action W[A] is gauge invariant, which implies that there is no "genuine" anomaly; the anomalous contributions from the fermion sector and from the Wess-Zumino boson sector cancel each other.

As promised before, I now come to the justification of Eq. (2.1), i.e. why the gauge field integral has to be performed over the complete configuration space. There are at least three arguments for this:

(i) Eq. (2.1) can be rewritten according to:

$$Z = \int dA \, e^{i\tilde{W}[A]}. \qquad (2.22)$$

In the anomalous case $\tilde{W}[A]$ is not gauge invariant; this means that the integration over different gauge fields, which are related by gauge transformations, does not give the same result. Hence, if gauge fixing were to be allowed, the result for Z would depend on the particular gauge which has been chosen for the calculation. This clearly contradicts the philosophy of gauge fixing: Nature does not care about a theorist's choice of gauge.

(ii) The usual argument as to why integration should only be performed over a gauge surface is not valid for anomalous theories. This argument goes as follows: The path integral has to be performed over all independent phase space variables. So the prescription is: first the constraints have to be analysed in order to find the independent variables, then the path integral is written down in the phase space representation (i.e. with the action in the Hamilton formalism $S = p\dot{q} - H$) and finally the momenta are integrated out. For the gauge invariant (non-anomalous) case this results in Eq. (2.13), i.e. integration only over a gauge surface. In the anomalous case, however, the first step is not possible, because the symmetry of the action does not represent the symmetry of the theory, since the fermionic measure is not gauge invariant. This feature can lead to first class constraints for a gauge non-invariant theory (e.g. in the conventional treatment of anomalous gauge theories) or to second class constraints for a gauge invariant theory (e.g. in the theory defined by Eq. (2.9)); neither of these results is correct. Hence the usual prescription does not provide any guidance: in the anomaly-free case integrating over (a) only independent degrees of freedom or (b) the entire configuration space, are equivalent (the result differs only by a constant factor), and in the anomalous case counting independent degrees of freedom is not possible. So from this point of view no preference can be given, either to Eq. (2.9) or to Eq. (2.13). Since Eq. (2.13) has proved to lead to inconsistencies it is natural to try what happens if path integration is performed over the complete configuration space.

(iii) In the chiral Schwinger model the explicit solution of Jackiw and Rajaraman[3] shows that there may be one more independent degree of freedom than expected from the non-anomalous case involving the ordinary Schwinger model[24]. This means that integration only over a gauge surface is not sufficient; there is one more degree of freedom which completes the integral over the entire configuration space. Hence in this specific model (which is the only one with an explicit solution), Eq. (2.9) turns out to be correct and Eq. (2.13) is incorrect.

Finally, I want to mention that the gauge fixing condition can be generalized to depend not only on the gauge field, but also on the scalar g[20]. Then the generating functional reads:

$$Z = \int dA\, dg\, d\psi\, d\bar{\Psi}\, \delta(f(A,g))\, \Delta_f[A,g]\, e^{iS_{st}[\bar{\Psi},\psi,A,g]} . \qquad (2.23)$$

To conclude this section, the essential statements to be kept in mind are: The correct quantization procedure for the gauge field automatically leads to a gauge (BRST) invariant quantum theory. Consequently there is no anomaly and the inconsistency in the equation of motion for the gauge field does not exist. Furthermore, BRST invariance leads us to expect ordinary (non-anomalous) Ward identities, which means that the usual argument against renormalizability is no longer valid. Certainly this is not sufficient to guarantee a sensible quantum theory, but the chance is better than ever before.

3. THE CHIRAL SCHWINGER MODEL

The motivations for considering the chiral Schwinger model are:

(i) The chiral Schwinger model is (1+1)-dimensional and in two dimensions the fermion integral can be performed explicitly. Hence it is possible to circumvent the intractable fermionic measure. Furthermore, the model is explicitly solvable, providing us with an excellent playground for testing the survival of the gauge symmetry, the anomaly cancellation and the consistency.

(ii) The model has been proven to be consistent, it was the counterexample against inconsistency of anomalous theories mentioned in the Introduction. However, consistency has been shown in a framework which does not reflect gauge invariance. Indeed, the mechanism of vector-boson mass-generation has been called an "anomalous breakdown of the gauge symmetry". The purpose of a recent investigation of G. Kramer and the present author[19] was to see what these results correspond to in a gauge invariant formulation of the model.

(iii) There is also a very practical reason for presenting the chiral Schwinger model in this talk. Namely, no other anomalous model has been treated in this framework up to now, so there is no alternative available.

Classically, the chiral Schwinger model is defined by the classical action

$$S[\bar{\psi},\psi,A] = \int \{-\tfrac{1}{4} F_{\mu\nu} F^{\mu\nu} + \bar{\psi}\gamma^{\mu}[i\partial_{\mu} + e\sqrt{\pi} A_{\mu}(1 + i\gamma_5)]\psi \} d^2x . \quad (3.1)$$

Here we use two dimensional notation:

$$\eta_{00} = -\eta_{11} = 1 \quad ; \quad \epsilon^{01} = -\epsilon_{01} = 1$$
$$\gamma_5 = i\,\gamma^0\,\gamma^1 \quad \Rightarrow \quad i\gamma_{\mu}\gamma_5 = \epsilon_{\mu\nu}\,\gamma^{\nu} . \quad (3.2)$$

The fermion integration can be done explicitly[3,25,26], here I only want to state the result (we ignore topologically nontrivial gauge configurations):

$$\tilde{W}[A] = \int \{-\tfrac{1}{4} F_{\mu\nu} F^{\mu\nu} + \frac{e^2}{2} A_{\mu} [\eta^{\mu\beta} a - (\eta + \epsilon)^{\mu\nu} \frac{\partial_{\nu}\partial_{\alpha}}{\Box} (\eta-\epsilon)^{\alpha\beta}]A_{\beta}\} d^2x. \quad (3.3)$$

This defines the chiral Schwinger model at the quantum level. However, the theory is completely determined only if, in addition to the classical action, it is stated how the fermion integration is to be regularized. This ambiguity is reflected in the arbitrariness of the regularization parameter a. Indeed, there has been some discussion about this parameter in the literature[27,28], but this has been settled and it is now clear that it is in fact necessary (Prof. Leutwyler explained in his lecture that $\ln \det D_L$ is well defined only up to a local polynomial. In our case it is just the A^2 term which is not fixed).

According to the results of the preceeding section there are two possibilities for quantization:

(i) start with Eq.(2.1): $$Z = \int dA \, e^{i\tilde{W}[A]} \, . \qquad (3.4)$$

(ii) start with Eq. (2.11): $$Z = \int \mathcal{D}A \, dg \, e^{i\tilde{W}[A^{g^{-1}}]} \, . \qquad (3.5)$$

In (i), manifest gauge invariance is lost and there seems to be an anomaly. Hence we call it the "anomalous" formulation. This is the formulation which has implicitly been used by Jackiw and Rajaraman[3] and in the Hamiltonian approach of Girotti, Rothe and Rothe[14]. In (ii), gauge invariance is manifest, thus we call it the "gauge invariant" formulation. This approach has been adopted in the Lagrange formalism of Refs. [9,16,20]. In the literature[17,18] there also exists a kind of "hybrid" formulation, starting from

$$Z' = \int \mathcal{D}A \, e^{i\tilde{W}[A]} \, . \qquad (3.6)$$

This amounts to fixing a gauge in a theory without gauge invariance. It leads to a violation of Lorentz invariance and is certainly incorrect.

In this talk I want to present a canonical quantization scheme based on the gauge invariant approach, in order to investigate the constraint structure, to show gauge invariance explicitly and to count the independent degrees of freedom properly. The resulting Hamiltonian quantum system will be compared with that following from the anomalous approach, which has been derived in Ref. [14].

If we parametrize g according to $g = e^{i\theta}$, then

$$A_\mu^{g^{-1}} = A_\mu - \frac{1}{e} \partial_\mu \theta$$

and the effective action reads:

$$\tilde{W}[A^{g^{-1}}] = \int \{ -\tfrac{1}{4} F_{\mu\nu} F^{\mu\nu} - \tfrac{1}{2}(a-1) \, \theta\Box\theta + e\theta\partial_\mu \, [(a-1)\eta^{\mu\nu} + \epsilon^{\mu\nu}] \, A_\nu$$

$$+ \frac{e^2}{2} A_\mu [\eta^{\mu\beta} a - (\eta+\epsilon)^{\mu\nu} \frac{\partial_\nu \partial_\alpha}{\Box} (\eta-\epsilon)^{\alpha\beta}] \, A_\beta \} \, d^2x \, . \qquad (3.7)$$

The nonlocal term $A \frac{\partial\partial}{\Box} A$ may be removed by an additional field Φ[3], this results in

$$Z = \int \mathcal{D}A\, d\Phi\, d\theta\, e^{i\int \mathcal{L}_{eff}\, d^2x} ,$$

$$\mathcal{L}_{eff} = -\tfrac{1}{4} F_{\mu\nu} F^{\mu\nu} - \tfrac{1}{2}\, \theta\Box\theta + e(\partial_\mu \Phi)(\eta - \epsilon)^{\mu\nu} A_\nu - \tfrac{1}{2}(a-1)\theta\Box\theta$$

$$- e(\partial_\mu\theta)[(a-1)\eta^{\mu\nu} + \epsilon^{\mu\nu}]A_\nu + \frac{e^2}{2} a A_\mu A^\mu . \tag{3.8}$$

Due to the fact that the noninvariant fermion measure has been eliminated, $\mathcal{L}_{eff}$ has to respect the symmetry of the theory, i.e. it has to be gauge invariant. Indeed, the gauge transformation

$$A_\mu \to A_\mu - \frac{1}{e}\partial_\mu \Lambda ; \quad \Phi \to \Phi + \Lambda ; \quad \theta \to \theta - \Lambda \tag{3.9}$$

leaves $\mathcal{L}_{eff}$ unchanged (up to a total divergence), which proves the manifest gauge invariance of the "gauge invariant" formulation.

Eq. (3.8) will be our starting point for canonical quantization. We expect first class constraints associated with the gauge invariance, contrary to the case of the "anomalous" formulation, where due to the lack of gauge invariance all constraints are second class. We proceed by calculating the canonical momenta; they are

$$\Pi_0 \approx 0 \tag{3.10}$$

$$\Pi_1 = -F_{01} = \partial_1 A^0 + \partial_0 A^1 \tag{3.11}$$

$$\Pi_\Phi = \partial_0 \Phi + e\,(A^0 + A^1) \tag{3.12}$$

$$\Pi_\theta = (a-1)\, \partial_0\theta - (a-1)\, e\, A^0 + e\, A^1. \tag{3.13}$$

Eq. (3.10) has to be construed as a constraint ($\approx$ denotes weak equality in the sense of Dirac[29]). In the case $a=1$ Eq. (3.13) is a constraint too, since it does not involve any velocity. Hence we have to distinguish between the cases $a \neq 1$ and $a = 1$. In the sequel I shall follow the procedure I indicated in my preceeding talk (these Proceedings), concerning Dirac's constraint analysis.

(a) The Case $a \neq 1$

In this case Eqs.(3.11)-(3.13) may be converted to express the velocities (except $\partial_0 A^0$) in terms of the momenta, in order to perform the Legendre transformation. Then the Hamiltonian density can be calculated to be:

$$\mathcal{H} = \tfrac{1}{2}\Pi_1\Pi_1 + \tfrac{1}{2}\Pi_\Phi^2 + \frac{1}{2(a-1)}\Pi_\theta^2 + A^0\partial\Pi_1 - e\Pi_\Phi(A^0 + A^1)$$

$$+ e\Pi_\theta(A^0 - \frac{1}{a-1}A^1) + \tfrac{1}{2}(\partial\Phi)^2 + \tfrac{1}{2}(a-1)(\partial\theta)^2 - e(\partial\Phi)(A^0 + A^1)$$

$$+ e(\partial\theta)[(a-1)A^1 - A^0] + \frac{e^2a^2}{2(a-1)}A^1A^1 + \xi_1\Pi_0 \ , \tag{3.14}$$

where ∂ denotes ∂_1. The Lagrange multiplier field ξ_1 reflects the fact that the velocity of A^0 is not fixed. The consistency requirement that the constraint in Eq.(3.10) be stable in time leads to Gauss' law ($H = \int \mathcal{H}\, dx$):

$$\dot{\Pi}_0 \approx -\frac{\delta H}{\delta A^0} = -(\partial\Pi_1 - e\Pi_\Phi + e\Pi_\theta - e\partial\Phi - e\partial\theta) \approx 0 \ . \tag{3.15}$$

Since the Poisson bracket of this expression with the Hamiltonian vanishes, we have two constraints;

$$\chi_1 = \Pi_0 \approx 0 \quad , \tag{3.16}$$

$$\chi_2 = \partial\Pi_1 - e\Pi_\Phi + e\Pi_\theta - e\partial\Phi - e\partial\theta \approx 0 \ . \tag{3.17}$$

Because of $\{\chi_1,\chi_2\} = 0$ the constraints are first class, as expected due to gauge invariance. According to the rules χ_2 must be added to the Hamiltonian with another Lagrange multiplier field ξ_2, leading to the total Hamiltonian

$$H_T = H + \int \xi_2\chi_2\, dx. \tag{3.18}$$

It is possible to fix the dynamics of the system, i.e. to eliminate the arbitrary Lagrange multiplier fields ξ_1 and ξ_2, by imposing two gauge conditions. Since we intend to compare our results with the anomalous approach which does not contain θ, we choose as our gauge conditions

$$\chi_3 = -\partial\theta \approx 0 \tag{3.19}$$

$$\chi_4 = \Pi_\theta - eA^1 + e(a-1)A^0 \approx 0 \ . \tag{3.20}$$

This corresponds to the gauge $\partial^\mu\theta = 0$ in the Lagrange formalism, which converts $\tilde{W}[A^{g^{-1}}]$ of the gauge invariant approach to $\tilde{W}[A]$ of the anomalous formulation. Hence these approaches coincide in the Lagrange formalism, up to the equation of motion with respect to θ, which is present only in the gauge invariant formulation. Also in the Hamilton formalism the two approaches are quite similar; if we insert χ_3 and χ_4 into χ_2 and $\mathcal{H}$, we find precisely the constraints and Hamiltonian of Ref.[14];

$$\chi_1 = \Pi_0 \approx 0 \, ,$$

$$\chi_2' = \partial\Pi_1 - e\Pi_\phi - e\partial\phi + e^2A^1 - e^2(a-1)\, A^0 \approx 0 \, , \tag{3.21}$$

$$\mathcal{H} \approx \tfrac{1}{2}(\pi_1)^2 + A^0\partial\Pi_1 + \tfrac{1}{2}[\Pi_\phi - e(A^0 + A^1)]^2 - \tfrac{1}{2}e^2aA_\mu A^\mu + \tfrac{1}{2}(\partial\phi)^2 - e(\partial\phi)(A^0 + A^1) \, . \tag{3.22}$$

The difference between the anomalous and the gauge invariant formulations in this gauge is the occurence of the additional constraints χ_3 and χ_4, which serve to eliminate θ and Π_θ. Now we have to check whether our gauge is an allowed gauge, i.e. whether it is achievable and complete. As may be seen from the transformation property of θ, Eq.(3.9), $\Lambda = \theta$ + constant converts an arbitrary θ to a constant field, which satisfies the gauge condition, hence the gauge is achievable. In order to see that the gauge is completely fixed we have to examine the constraint algebra. The equal time Poisson brackets form the matrix:

$$\{\chi(x),\chi(y)\} = \begin{bmatrix} 0 & 0 & 0 & -e(a-1) \\ 0 & 0 & -e\partial_x & 0 \\ 0 & -e\partial_x & \partial_x & -\partial_x \\ e(a-1) & 0 & -\partial_x & 0 \end{bmatrix} \delta(x-y) \tag{3.23}$$

where x and y denote space coordinates only. The determinant of $\{\chi,\chi\}$ does not vanish as long as $a \neq 1$, therefore the gauge is completely fixed. For the purpose of quantization we need the Dirac brackets, because these transform to the commutators of the corresponding quantum fields. To this aim $\{\chi,\chi\}$ has to be inverted:

$$C(x,y) = \begin{bmatrix} 0 & \frac{-1}{e^2(a-1)} & 0 & \frac{1}{e(a-1)} \\ \frac{1}{e^2(a-1)} & 0 & \frac{-1}{e\partial_x} & 0 \\ 0 & \frac{-1}{e\partial_x} & 0 & 0 \\ \frac{-1}{e(a-1)} & 0 & 0 & 0 \end{bmatrix} \delta(x-y) \tag{3.24}$$

such that

$$\int C_{ij}(x,y)\, \{\chi_j(y),\chi_k(z)\}\, dy = \delta_{ik}\delta(x-z) \, . \tag{3.25}$$

From Eq.(3.24) the calculation of the Dirac brackets is straightforward, they read:

$$\{A^0(x),A^1(y)\}_D = \frac{-1}{e^2(a-1)} \partial_x\delta(x-y) \; ; \quad \{A^1(x),\Pi_\theta(y)\}_D = \frac{1}{e} \partial_x\delta(x-y)$$

$$\{A^0(x),\phi(y)\}_D = \frac{1}{e(a-1)} \delta(x-y) \quad ; \quad \{\Phi(x),\Pi_\theta(y)\}_D = \delta(x-y)$$

$$\{A^0(x),\Pi_1(y)\}_D = \frac{1}{a-1} \delta(x-y) \quad ; \quad \{A^0(x),\Pi_\theta(y)\}_D = \frac{-1}{e(a-1)} \partial_x\delta(x-y)$$

$$\{A^0(x),\Pi_\Phi(y)\}_D = \frac{-1}{e(a-1)} \partial_x \delta(x-y) \quad ; \quad \{\Pi_\Phi(x),\Pi_\theta(y)\}_D = \partial_x\delta(x-y)$$

$$\{A^1(x),\pi_1(y)\}_D = \delta(x-y) \quad , \quad \{\Phi(x),\pi_\Phi(y)\}_D = \delta(x-y) \tag{3.26}$$

and all other Dirac brackets vanish. The corresponding commutators coincide completely with those of the anomalous approach[14], except for those involving Π_θ, which merely express the fact that Π_θ is a dependent quantity. Since also the Hamiltonian of Eq. (3.22) is the same as in Ref.[14], this gauge precisely reproduces the quantum system defined by the anomalous approach to the chiral Schwinger model. This once more establishes the equivalence of the generating functionals of Eqs.(2.1) and (2.9) in this explicit example: using the anomalous formulation is nothing else than working in a specific gauge of the gauge invariant formulation. This statement is also true for the case $a = 1$, as we shall see below.

The quantum system may be formulated in the following (unconstrained) way:

$$H = \int \{ \tfrac{1}{2} \Pi_1{}^2 + \tfrac{1}{2}[\Pi_\Phi - eA^1]^2 + \tfrac{1}{2}[\partial\Phi - eA^1]^2 + \tfrac{1}{2}(a-1)e^2(A^1)^2 + \tfrac{1}{2}(a-1)(A^0)^2_{symm} \} \, dx, \tag{3.27}$$

where A^0 is shorthand for

$$A^0 = \frac{1}{e^2(a-1)} (\partial\Pi_1 - e\Pi_\Phi - e\partial\Phi + eA^1), \tag{3.28}$$

and the canonical equal-time-commutators are

$$[\Phi(x),\Pi_\Phi(y)] = [A^1(x),\Pi_1(y)] = i\delta(x-y). \tag{3.29}$$

The last term in H in Eq. (3.27) has to be symmetrized according to the general rules of nonlinear quantum mechanics[30,31], since A^0 contains noncommutative quantities.

The established equivalence of the gauge invariant and the anomalous formulation implies that the results of the latter are valid in our case too[14]: for $a > 1$ the quantum system is consistent and unitary, containing a massive ($m^2 = \frac{e^2a^2}{a-1}$) and a massless degree of freedom; for $0<a<1$ or $a<0$ the theory contains a tachyon ($m^2<0$); for $a=0$ the theory is inconsistent, since the solution of the Heisenberg equations of motion is not compatible with the commutator structure. It is also clear from the Hamiltonian in Eq. (3.27) that a sensible quantum theory can be expected only as long as $a>1$, since otherwise the potential is not bounded from below and the vacuum could contain infinitely large fields.

(b) The Case $a = 1$

I wish to discuss this case only very briefly, since the procedure is essentially the same as in the case $a \neq 1$. The Lagrange density simplifies to (Cf. Eq. (3.8))

$$\mathcal{L}_{eff} = -\tfrac{1}{4}F_{\mu\nu}F^{\mu\nu} - \tfrac{1}{2}\Phi\Box\Phi + e(\partial_\mu\Phi)(\eta-\epsilon)^{\mu\nu}A_\nu + \tfrac{1}{2}e^2A_\mu A^\mu - e(\partial_\mu\theta)\epsilon^{\mu\nu}A_\nu). \quad (3.30)$$

The canonical momenta are

$$\Pi_0 = 0\ , \qquad \Pi_\theta = eA^1 \quad (3.31)$$

$$\Pi_1 = -F_{01}\ , \ \Pi_\Phi = \partial_0\Phi + e(A^0 + A^1). \quad (3.32)$$

Eqs. (3.31) are constraints since they do not involve velocities, consistency requires that they be stable in time. This leads via Poison brackets with the Hamiltonian to two more constraints. These four constraints can be rearranged in such a way that two of them are first class and the others second class. Again, the presence of first class constraints reflects gauge invariance. This means that we have to impose gauge conditions; we use

$$\partial\theta \approx 0\ ; \quad \Pi_\theta - \Pi_\Phi - \partial\Phi + e(A^0 + A^1) \approx 0. \quad (3.33)$$

By linear combination of the four constraints and the gauge fixing conditions we define the following set of constraints:

$$\begin{aligned} \chi_1 &= \Pi_0 \approx 0 \ ; & \chi_2 &= -\partial\Pi_1 + e\Pi_\Phi + e\partial\Phi - e^2A^1 \approx 0, \\ \chi_3 &= \Pi_1 \approx 0 \ ; & \chi_4 &= e(A^0 + 2A^1) - \Pi_\Phi - \partial\Phi \approx 0 \\ \chi_5 &= \partial\theta \approx 0 \ ; & \chi_6 &= \Pi_\theta - eA^1 . \end{aligned}$$

(3.34)

Again, the constraints independent of θ and π_θ coincide with those of the anomalous approach, χ_5 and χ_6 serve to eliminate θ and π_θ. The final task is the calculation of the commutators via Dirac brackets, the result is as follows:

$$[\Phi(x),\pi_\Phi(y)] = i\delta(x-y) \ , \tag{3.35}$$

all other commutators can be found by expressing all quantities in terms of Φ and π_Φ according to the constraints and then Eq. (3.35) may be used. With the constraints of Eq. (3.34) the Hamiltonian simplifies to

$$H = \int (\tfrac{1}{2}\,\pi_\Phi{}^2 + \tfrac{1}{2}(\partial\Phi)^2)\,dx \ , \tag{3.36}$$

and this, together with the commutators, just reproduces the quantum system of the anomalous approach. From Eqs. (3.35) and (3.36) it is clear that the theory is consistent and unitary, and that it contains one massless scalar field.

4. DISCUSSION

The Chiral Schwinger Model

We have shown explicitly that the anomalous formulation of the chiral Schwinger model is nothing else but a specific gauge of the gauge invariant formulation. This clarifies the reason for the absence of an anomaly in current conservation and current commutators in the anomalous approach. In order to appreciate the advantages of the gauge invariant formulation it is useful to contrast current conservation and current commutators in the two cases.

(i) the "anomalous" approach

Here the superficial anomalous divergence of the current reads[15]:

$$\partial_\mu j^\mu = e^2[(1-a)\partial_\mu A^\mu - \epsilon^{\mu\nu}\partial_\mu A_\nu] \ . \tag{4.1}$$

This is seen to vanish only after the theory is completely solved, i.e. if the solution of the Euler-Lagrange equations is used. This solution follows from $\tilde{W}[A]$ (Eq. (3.3)), with the nonlocal term eliminated by Φ:

$$A_\mu = \frac{-1}{ea}[\partial_\mu\Phi + (1-a)\epsilon_{\mu\nu}\partial^\nu\Phi - a\epsilon_{\mu\nu}\partial^\nu h] \ , \tag{4.2}$$

with h a harmonic function.

The charge density j_0 is given by

$$j_0 = e(\Pi_\Phi + \partial\Phi) - e^2A^1 + e^2(a-1)A^0 . \qquad (4.3)$$

This leads to an anomalous Schwinger term in the Poisson bracket

$$\{j_0(x),j_0(y)\} = 2e^2 \partial_x\delta(x-y) , \qquad (4.4)$$

which dissappears only at the level of the Dirac bracket (and therefore of quantum commutators).

(ii) the gauge invariant approach

The equation of motion for θ in the gauge $\theta=0$ reads (cf. Eq. (3.8))

$$\partial_\mu[(a-1)\eta^{\mu\nu} + \epsilon^{\mu\nu}]A_\nu = 0, \qquad (4.5)$$

which immediately implies that the current is conserved, compare Eq.(4.1). Here current conservation, which seems to be an accident in the anomalous approach, is ensured by gauge invariance. Furthermore, it is not necessary to have a complete solution of the theory available in order to establish current conservation. The charge density is given by

$$j_0 = e(\Pi_\Phi - \Pi_\theta + \partial\Phi + \partial\theta) , \qquad (4.6)$$

which implies that there is no Schwinger term even at the Poisson bracket level:

$$\{j_0(x),j_0(y)\} = 0. \qquad (4.7)$$

Hence we may conclude: it is just the introduction of the field θ, i.e. the requirement of gauge invariance, which makes the absence of genuine anomalies transparent. Hence it seems to be advantageous to use the gauge invariant formulation, not only in the chiral Schwinger model, which is only a toy model, but also in other models and higher dimensions.

Before leaving this model, I wish to make two remarks. (i) The appearance of a vector boson mass has nothing to do with an explicit breakdown of the gauge symmetry, since this feature occurs not only in the gauge invariant formulation of the chiral Schwinger model, but also in the ordinary Schwinger model where gauge invariance is not endangered at all. (ii) One might worry about the fact that there are regularization prescriptions which lead to inconsistent quantum theories ($a<1$). However, it has to be noted that a theory is considered to be consistent if there is at least one regularization which does not lead to inconsistencies. For instance, the ordinary Schwinger model, Eq. (3.1) without γ_5, essentially has the effective action of Eq. (3.3) without the ϵ's. This theory is consistent *only* if $a = 1$, i.e. if $\mathfrak{V}[A]$ is chosen to be gauge invariant. Therefore, we are accustomed to a situation where consistency depends on the regularization prescription, hence this feature should not be considered dangerous in the gauge invariant approach to "anomalous" gauge theories[32].

General Conclusions

There are two ways to quantize anomalous gauge theories; the anomalous formulation

$$Z = \int dA\, d\psi\, d\bar{\psi}\, e^{iS[\bar{\psi},\psi,A]} \tag{4.8}$$

and the gauge invariant formulation

$$Z = \int \mathcal{D}(A,g)\, d\psi\, d\bar{\psi}\, e^{i\{S[\bar{\psi},\psi,A] + \alpha_1[A,g^{-1}]\}}\ , \tag{4.9}$$

with

$$\mathcal{D}(A,g) = dA\, dg\, \delta(f(A,g))\, \Delta_f[A,g]\ . \tag{4.10}$$

The traditional approach

$$Z = \int \mathcal{D}A\, d\psi\, d\bar{\psi}\, e^{iS[\bar{\psi},\psi,A]}\ , \tag{4.11}$$

however, is incorrect if there is an anomaly in the fermion sector. Provided that quantization is performed correctly, gauge invariance remains preserved. Hence there is no anomaly, "anomalous" gauge theories are anomaly free. This in turn implies that the stated inconsistency in the classical equation of motion for the gauge field does not exist.

Finally, the gauge invariant approach is superior to the anomalous one, not only in the chiral Schwinger model, where we have demonstrated this explicitly, but also in other models. Especially if the dimension is higher than two (this covers all interesting cases), the anomalous formulation is not very useful since, for example, it does not allow us to perform perturbation theory, because it is impossible to define a propagator for the gauge field.

REFERENCES

1. H. Leutwyler, these Proceedings.
2. For a review, see R. Jackiw, in: "Relativity, Groups and Topology II" B. DeWitt and R. Stora, eds., North Holland, Amsterdam (1984).

3. R. Jackiw and R. Rajaraman, Phys. Rev. Lett. 54: 1219,2060(E) (1985).

4. E. D'Hoker and E. Farhi, Nucl Phys. B248: 59,77 (1984).

5. A. J. Niemi and G. W. Semenoff, Princeton preprint, September 1985.

6. L. D. Faddeev and S. L. Shatashvili, Phys. Lett. B167: 225 (1986).

7. O. Babelon, F. A. Shaposnik and C. M. Viallet, Phys.Lett. B177: 385(1986).

8. A. V. Kulikov, Serpukhov preprint IHEP 86-083.

9) K. Harada and I. Tsutsui, Phys. Lett. B183: 311 (1987).

10. L. D. Faddeev and V. N. Popov, Phys. Lett. B25: 29 (1967).

11. J. Wess and B. Zumino, Phys. Lett. B37: 95 (1971).

12. C. Becchi, A. Rouet and R. Stora, Phys. Lett. B52: 344 (1974); Comm. Math. Phys. 42: 127 (1975).

13. I. V. Tyutin, Int. report FIAN 39 (1975)

14. H. O. Girotto, H. J. Rothe and K. D. Rothe, Phys. Rev. D33: 514 (1986); D34: 592 (1986).

15. R. Rajaraman, Phys. Lett. B154: 305 (1985).

16. F. A. Schaposnik and J. N. Webb, Manchester preprint MC-TH-86-17.

17. M. Chanowitz, Phys. Lett. B171: 280 (1986).

18. J. G. Holliday, E. Rabinovici, A. Schwimmer and M. Chanowitz, Nucl. Phys. B268: 413 (1986).

19. N. K. Falck und G. Kramer, Hamburg preprint DESY 86-145 (1986), to appear in Ann. Phys. (N.Y.).

20. K. Harada and I. Tsutsui, Tokio preprints TIT-HEP 101,102 (1986).

21. C. A. Linhares, H. J. Rothe and K. D. Rothe, Heidelberg preprint HD-THEP-86-20.

22. A. J. Niemi and G. W. Semenoff, Phys. Lett. B175: 439 (1986).

23. K. Fujikawa, Phys. Rev. D21: 2848 (1980), D22:1499(E) (1980).

24. J. Schwinger, Phys. Rev. 128: 2425 (1962).

25. J. N. Webb, Z. Phys. C31: 295 (1986).

26. K. Harada, H. Kubota and I. Tsutsui, Phys. Lett. B173: 77 (1986).

27. C. Hagen, Phys. Rev. Lett. 55: 2223(C) (1985).

28. R. Jackiw and R. Rajaraman, Phys. Rev. Lett. 55: 2224(C) (1985).

29. P. A. M. Dirac, Can. J. Phys. 2: 125 (1950); "Lectures on Quantum Mechanics", Yeshiva Univ. Press, New York (1964).
30. N. K. Falck, these Proceedings.
31. B. S. DeWitt, Phys. Rev. 85: 653 (1952).
32. M. Omote and H. Sato, Prog. Theor. Phys. 47: 1367 (1972).
33. R. Jackiw, MIT preprint CTP 1436 (1986).
34. R. D. Ball, Phys. Lett. B183: 315 (1987).

ANOMALIES IN ODD DIMENSIONS

M. Reuter*

Deutsches Elektronen Synchrotron DESY
Notkestr. 85, 2000 Hamburg 52, Fed. Rep. Germany

ABSTRACT

Gauge theories with fermions in (2n+1)-dimensions show the phenomenon of fractional fermion number and for n=1 they are closely related to topologically massive gauge theories. These effects are related to an anomaly in the vacuum current. The anomaly is calculated by use of the zeta function method for regularizing the fermion determinant. The heat kernel method relates the anomaly to the spectral asymmetry of the Dirac Hamiltonian. The contribution of the anomalous part of the vacuum current to the effective action is shown to be given by the Chern-Simons term of the respective dimensionality. The relationship between different kinds of anomalies in even and odd dimensions is explored. Physical systems in which these anomalies manifest themselves involve cosmic strings and domain walls.

*Address after September 1987: CERN Theory Division, Geneva, Switzerland.

Gauge theories with fermions in (2n+1)-dimensions are interesting for at least two reasons: they show the phenomenon of a fractional fermion number (i.e. topologically non-trivial field configurations like vortices or monopoles can carry half-integer fermion number), and for $n = 1$ they are closely related to the topologically massive gauge theories discovered by Deser, Jackiw and Templeton[1,2].

In this Introduction we first illustrate the mechanism leading to a fractional fermion number by considering a simple example (a more detailed discussion has been given by Jackiw[3]). We consider D-dimensional fermions, D not necessarily odd, interacting with static external fields. Furthermore, we assume a linear relation between energy and momentuum leading to a Hamiltonian of the general form

$$H = \vec{\alpha}\cdot\vec{p} + \sum_n \beta_n \Phi_n(\vec{x}) , \quad \vec{p} = - i \vec{\nabla} . \tag{1}$$

Here α and β_n are constant hermitian matrices, and the Φ_n's are arbitrary (real) background fields. The spectrum of the fermions has to be determined from the eigenvalue equation

$$H\Psi_E = E\Psi_E . \tag{2}$$

It has the structure of the Dirac equation, but also in solid state physics equations like (2) can occur[4]. Assuming to have solved the eigenvalue problem (2) for a given set of Φ_n's the fermionic vacuum is constructed by occupying all states with $E < 0$ and leaving empty all states with $E > 0$. Thus the fermion density becomes

$$\int_{-\infty}^{0} dE \, \Psi_E^*(\vec{x}) \, \Psi_E(\vec{x}) .$$

This quantity is divergent. It can be renormalized by substracting the density for the vanishing background fields:

$$\rho(\vec{x}) = \int_{-\infty}^{0} dE \, \{ \Psi_E^*(\vec{x}) \, \Psi_E(\vec{x}) - \Psi_{OE}(\vec{x}) \, \Psi_{OE}(\vec{x}) \} . \tag{3}$$

The wave functions ψ_{OE} are solutions of (2) for $\Phi_n \equiv 0$. The charge induced in the fermion vacuum by the background fields reads

$$N = \int d^{D-1}x \, \rho(\vec{x}) \, . \tag{4}$$

If the Φ_n have a one-soliton profile, we identify N with the fermion number of this soliton. To show that N is not necessarily an integer we make the simplifying assumption that there exists a unitary conjugation matrix which anticommutes with the Hamiltonian:

$$\{H,C\} = 0. \tag{5}$$

This implies that the spectrum of H is symmetric with respect to E = 0, because applying C to ψ_E changes the sign of its eigenvalue:

$$H\,(C\psi_E) = -\,E\,(C\psi_E)\,, \quad C\psi_E = \psi_{-E}\,. \tag{6}$$

Moreover, we assume that the Φ_n's are such that there is at least one normalizable solution of (2) with E = 0 (typically the existence of such zero modes is related to the topology of the background field by index theorems[5]). Now the definition of the ground state becomes ambiguous, since it is not clear whether or not the E = 0 mode should be occupied. If there is only one normalizable solution $\Phi(x)$ with vanishing energy it turns out that the only consistent fermion number assignment is $\pm$ 1/2. To see this we substract the completeness relations of $\{\psi_E\}$ and $\{\psi_{OE}\}$ from each other to obtain

$$\int_{-\infty}^{\infty} dE \, \{\psi_E^*(\vec{x})\,\psi_E(\vec{x}) - \psi_{OE}^*(\vec{x})\,\psi_{OE}(\vec{x})\} = 0 \, .$$

Separating Φ and using (6) yields

$$2\int_{-\infty}^{0} dE \, \{\psi_E^*(\vec{x})\,\psi(\vec{x}) - \psi_{OE}^*(\vec{x})\,\psi_{OE}(\vec{x})\} + \Phi^*(\vec{x})\,\Phi(\vec{x}) = 0 \, ,$$

hence the charge density (3) reads

$$\rho(\vec{x}) = - \tfrac{1}{2} \Phi^*(\vec{x}) \Phi(\vec{x}) \quad ,$$

and the vacuum charge is $N = -\frac{1}{2}$, qed. The vacuum charge being half-integer is a direct consequence of the existence of the matrix C; in the most general case N may assume arbitrary real values. The lesson to be learned from this discussion is that non-integer fermion numbers can appear if the relevant Hamiltonian or Dirac operator allows for normalizable zero modes. In the following section we shall come back to this point in the context of (2n+1)-dimensional gauge theories.

As a further preparation we now briefly discuss the essential features of the topologically massive gauge theories[1,2]. Deser et al. discovered that in (2n+1)-dimensions it is possible to construct a gauge invariant mass term for Yang-Mills fields. The gauge group could be U(1) or SU(N); in the latter case the gauge invariant field equations for the Lie-algebra-valued field strength $F_{\mu\nu} = F^a_{\mu\nu} T^a$ read (T^a is in the fundamental representation)

$$D_\mu F^{\mu\nu} + \frac{\mu}{2} \epsilon^{\nu\alpha\beta} F_{\alpha\beta} = 0 \quad , \tag{8}$$

where μ is an arbitrary mass parameter. An equivalent form is

$$[D_\alpha D^\alpha + \mu^2] \, {}^*F_\mu = i \, \epsilon_{\mu\alpha\beta} \, [{}^*F^\alpha , {}^*F^\beta] \tag{9}$$

with

$${}^*F_\mu := \tfrac{1}{2} \epsilon_{\mu\alpha\beta} F^{\alpha\beta} \, . \tag{10}$$

It can be shown that (8) or (9) indeed describe vector bosons of mass μ. The classical action from which these equations of motion follow is given by

$$I[A] = - \frac{1}{2g^2} \int d^3x \; tr \; (F_{\mu\nu} F^{\mu\nu}) + \frac{8\pi\mu^2}{g^2} W[A] \; , \tag{11}$$

with the Chern-Simons term

$$W[A] = \frac{1}{8\pi^2} \epsilon^{\alpha\beta\gamma} \int d^3x \; tr \; (A_\alpha \partial_\beta A_\gamma + \frac{2}{3} i \, A_\alpha A_\beta A_\gamma) \, . \tag{12}$$

As is well known from the standard instanton discussion, the Chern-Simons term changes under "large", i.e. topologically non-trivial, gauge transformations U by the winding number w[U] of U :

$$W[A^U] = W[A] + w[U] . \tag{13}$$

A gauge transformation $U = U(x)$ can be considered a map from the compactified space-time, which is taken to be a large S^3, into the gauge group. The space of these maps consists of equivalence classes of mutually homotopic gauge transformations and has the group structure $\Pi_3(SU(N)) = \mathbb{Z}$. The integer w[U] tells us to which class the transformation U belongs. This means that despite the equations of motion being gauge invariant, the classical action I[A] is not invariant under all gauge transformations, but can change by integer multiples of $8\pi^2\mu/g^2$. For the quantum theory based upon (11) to be well defined, it is necessary that the factor exp(iI[A]) appearing in the path integral be a gauge invariant functional of A_μ. This forces the constraint

$$\frac{4\pi\mu}{g^2} = n \in \mathbb{Z} \tag{14}$$

on the parameters of the theory, guaranteeing that I[A] changes only by irrelevant multiplies of 2π. This quantization condition is very similar to the quantization of the electric and magnetic charge in the theory of the Dirac monopole. In both cases the classical equations of motion are gauge invariant. In the quantum theory, however, one has to use the underlying action, which is not gauge invariant. Its gauge dependence has no consequences only if the parameters of the theory fulfill certain quantization conditions. For further details we refer to Refs.[7] and [8].

In the abelian analogue of (8) no parameter quantization occurs. Because $\Pi_3(U(1))$ is trivial, all gauge transformations are homotopic and I[A] is invariant.

Another interesting property of these 3-dimensional models was first discussed by Redlich[9]. He considered ordinary SU(N) gauge fields coupled to massless Dirac fermions in the fundamental representation of SU(N). The classical action of this system is invariant under gauge as well as parity transformations (in 3 dimensions "parity" means $(t,x,y) \to (t,-x,y)$). However, integrating out the fermions one finds that this is no longer true for the effective action

$$I_{eff}[A] = -i \ln \det [i\not{D}(A)] \quad . \tag{15}$$

It turns out that it is not possible to regularize the fermionic determinant in a way which respects parity and gauge invariance simultaneously. Keeping parity as an intact symmetry at the quantum level, Redlich found that the determinant unavoidably changes its sign under gauge transformations with an odd winding number:

$$\det [i\not{D}(A^U)] = (-1)^{w[U]} \det [i\not{D}(A)] \quad .$$

This means that $I_{eff}[A]$ can change by odd multiples of π. On the other hand, one can use the gauge invariant Pauli-Villars scheme. In this case parity is spoiled from the outset because in three dimensions a fermion mass term $M\bar{\psi}\psi$ is P-violating. The regularized effective action is obtained by subtracting the contribution of the regulator field for its mass M going to infinity:

$$I^R_{eff} [A] = I_{eff} [A] - \lim_{M\to\infty} I_{eff} [A,M] \, . \tag{16}$$

The second part of (16) contains the term $\pi W[A]$ which changes under gauge transformations by $\pi w[U]$, and therefore compensates for the gauge non-invariance of the first piece $I_{eff}[A]$. However, $W[A]$ is easily seen to be P-violating (this situation is similar to the conflict between chiral symmetry and gauge invariance in 4 dimensions). The renormalized effective action (16) is of the general form

$$I^R_{eff}[A] = I_{NA}[A] - \frac{M}{|M|} \pi \, W[A] \, . \tag{17}$$

The first part I_{NA} is a complicated non-analytical functional of A_μ, but parity conserving. The only parity violation comes from $W[A]$. Note that even for $M \to \infty$ the second term depends on the sign of the regulator mass (in the corresponding U(1)-gauge theory, QED_3, the factor of π in (17) is replaced by 2π). The result (17) shows that ordinary gauge fields interacting with (massless) fermions dynamically generate a topological mass term à la Deser et al. In the following sections we shall discuss this mechanism in some detail.

We are now going to show that for a (2n+1)-dimensional (euclidian) theory of fermions interacting with classical gauge fields, the vacuum current contains an anomalous piece corresponding to an effective action which is precisely the Chern-Simons term of the respective dimensionality. For n = 1 we recover the results above. We shall use massive fermions; this allows a gauge invariant regularization of the fermion determinant, but sacrifices parity. Following an idea of Niemi and Semenoff[10], we shall relate this anomalous current to the (non-abelian) chiral anomaly in 2n-dimensions. We perform the calculations for an SU(N) gauge group with the fermions transforming according to its fundamental representation, but with some minor modifications the results also apply to the U(1) case.

We start from the Lagrangian

$$\mathcal{L} = \bar{\psi}(i\not{D} - m)\psi \equiv \bar{\psi}(\, i\gamma^{\mu}\{\partial_{\mu} + iA^{a}_{\mu}T^{a}\} - m\,)\psi\ .$$

From now on we are working in (2n+1)-dimensional euclidian space with metric $g_{\mu\nu} = -\delta_{\mu\nu}$ (Greek indices run from 0 to 2n, whereas latin ones run from 1 to 2n). The Dirac matrices are taken to be antihermitian. We define the following generating functional

$$Z[A] = \int [d\psi\, d\bar{\psi}]\, \exp\, \{\, \int d^{2n+1}x\ \bar{\psi}(i\not{D}-m)\psi\, \} = \det\, [i\not{D} - m]\ . \tag{18}$$

The ("charged") vacuum current

$$j^{\mu a}(x) = \frac{\langle 0\,|\bar{\psi}(x)\gamma^{\mu}T^{a}\psi(x)\,|0\rangle^{A}}{\langle 0\,|0\rangle^{A}} \tag{19}$$

can be expressed as

$$j^{\mu a}(x) = -\,\frac{1}{Z[A]}\,\frac{\delta Z[A]}{\delta A^{a}_{\mu}(x)} = -\,\frac{\delta}{\delta A^{a}_{\mu}(x)}\,\ln\det\,[i\not{D} - m]\ . \tag{20}$$

To explicitly evaluate (20) we assume a static (i.e. x^{0}-independent) gauge potential with $A_0 = 0$, i.e. a purely magnetic field. Furthermore, we only calculate the 0-component of j^μ and deduce the other components by covariance arguments[10]. Let us introduce

$$\tilde{\not{D}} = \gamma^0\partial_0 + \not{D}_{2n}$$
$$\not{D}_{2n} = \gamma^k(\partial_k + iA_k^a T^a) , \quad k = 1,2,\dots,2n , \tag{21}$$

where $\not{D}_{2n}$ depends only on the 2n variables $\vec{x} \equiv (x^i)$. Hence one has

$$j^{0a}(x) = - \frac{\delta}{\delta A_0^a(x)} \ln \det [- \gamma^0 A_0^b T^b + i\tilde{\not{D}} - m] \Big|_{A_0=0} . \tag{22}$$

The determinant in (22) cannot be naively defined as a product of eigenvalues, because the spectrum of the operator in the square bracket is unbounded from above (for A_μ=0 the square of the operator has eigenvalues p^2+m^2!). To give a well-defined meaning to this determinant we use the zeta-function method[11]. To see how it works, let us consider an arbitrary positive operator Ω which has a complete set of eigenfunctions:

$$\Omega_x f_m(x) = \lambda_m f_m(x).$$

The associated zeta-function is defined by

$$\zeta(\Omega|s) := \sum_m \lambda_m^{-s} , \quad s \in \mathbb{C} . \tag{23}$$

It can be shown that this sum converges for the real part of s sufficiently large, and that the complex function thus defined can be analytically continued to a meromorphic function on the entire s-plane, which is non-singular at s=0 for a large class of operators. The regularized determinant of Ω is defined as

$$\det \Omega := \exp\{ - \frac{d}{ds} \zeta(\Omega|s)|_{s\equiv 0}\} , \tag{24}$$

with an analytical continuation to s=0 understood. It is easy to see that for finite matrices (where it is possible to interchange the differentiation in (24) with the sum of (23)) this definition reduces to the usual product of eigenvalues.

In many applications it is not possible to explicitly determine the spectrum of Ω and to do the sum of (23). In particular, if one is not interested in $\det \Omega$ itself but only in its variation with respect to some parameter, the following procedure is more convenient. One determines the "heat-kernel"

$$G(x,y;t) = \sum_m e^{-\lambda_m t} f_m(x) f_m^+(y), \quad t > 0 \tag{25}$$

as the solution of the differential equation

$$(\Omega_x + \frac{\partial}{\partial t}) G(x,y;t) = 0 \tag{26a}$$

with the initial condition

$$G(x,y;0) = \delta(x-y) . \tag{26b}$$

Then the zeta-function is given by a Mellin transformation:

$$\zeta(\Omega|s) = \frac{1}{\Gamma(s)} \int_0^\infty dt\, t^{s-1} \int dx \,\mathrm{tr}\, G(x,x;t) . \tag{27}$$

The trace refers to possible algebraic (spinor, color, ...) indices. Eq. (27) is a simple consequence of the Euler representation of the Γ-function.

The definition (24) can not yet be applied to (22) because the operator in question is not positive definite. We therefore add a suitable constant (with respect to A_0) to (24) :

$$\begin{aligned} j^{0a}(x) &= - \frac{\delta}{\delta A_0^a(x)} \{ \ln \det [- \gamma^0 A_0^b T^b + i\tilde{\not\partial} - m] + \ln \det [- i\not\partial - m] \} \Big|_{A_0=0} \\ &= - \frac{\delta}{\delta A_0^a(x)} \{ \ln \det [- \gamma^0 A_0^b T^b (i\not\partial + m) + \tilde{\not\partial}^2 + m^2] \Big|_{A_0=0} \end{aligned} \tag{28}$$

For $A_0 = 0$ the new operator reduces to $\not\partial^2 + m^2$ which (in euclidian space) is hermitian and positive. Because we may consider A_0 infinitesimal, it is easy to convince oneself that the A_0-term does not spoil the applicability of the heat-kernel method. The next step is to define the determinant in the last line of (28), via Eqs. (24) and (27), and to calculate the functional derivative of the zeta-function. For the details of this slightly lengthy calculation we refer to Refs. [12] and [13]. The final answer reads

$$j^{0a}(x) = -m \int_{-\infty}^{\infty} \frac{dk_0}{2\pi} \frac{d}{ds}\bigg|_0 \frac{s}{\Gamma(1+s)} \int_0^{\infty} dt\, t^s \exp\left[-(m^2 + k_0^2)t\right]$$

$$\cdot \lim_{\vec{x}' \to \vec{x}} \operatorname{tr} \left[\gamma^0 T^a e^{-\not{D}_{2n}^2 t}\right]_{\vec{x}} \delta^{(2n)}(\vec{x} - \vec{x}') . \qquad (29)$$

For the corresponding expression of the singlet current (where the generator T^α of Eq. (19) is omitted) one simply has to omit T^α from (29). To interpret Eq. (29) the following observation is crucial: if we consider a 2n-dimensional theory of fermions $\chi(\vec{x})$, with mass $M = (m^2+k_0^2)^{\frac{1}{2}}$, which interact with the spatial part $A_i^a(\vec{x})$ of our gauge field, and calculate within this theory the vacuum expectation value $\langle\bar{\chi}(\vec{x})\gamma^0 T^a \chi(\vec{x})\rangle^A$, the result is almost the same as (29):

$$j^{0a}(x) = \int_{-\infty}^{\infty} \frac{dk_0}{2\pi} \frac{m}{M} \langle\bar{\chi}(\vec{x})\gamma^0 T^a \chi(\vec{x})\rangle_M^A . \qquad (30)$$

To prove (30) one introduces a generating functional similar to (18) and proceeds as above. In this 2n-dimensional auxiliary theory (which is only a mathematical tool) the matrix γ_0 plays the same role as γ_5 does in 4 dimensions: it anticommutes with all Dirac matrices contained in $\not{D}_{2n}$ and it defines the chirality of the χ-fermions . This enables us to calculate a part of the terms appearing in (30) by exploiting the non-abelian chiral anomaly[14] in 2n dimensions:

$$\partial_k \langle\bar{\chi}\gamma^k\gamma^0 T^a \chi\rangle + i\langle\bar{\chi}[T^a, A]\gamma^0\chi\rangle = 2iM \langle\bar{\chi}\gamma^0 T^a \chi\rangle + i\mathcal{A}_{2n}^a . \qquad (31)$$

The anomaly factor reads

$$\mathcal{A}_{2n}^a = K_n\, \epsilon_{k_1 k_2 \ldots k_{2n}} \operatorname{tr}(T^a F^{k_1 k_2} \ldots F^{k_{2n-1} k_{2n}}) , \qquad (32)$$

where $K_n = (-1)^n (2^{2n-1}\pi^n n!)^{-1}$. Obviously, the first term on the RHS of (31) has the same form as the integrand in Eq. (30), which may therefore be replaced by

$$j^{0a}(x) = \int_{-\infty}^{\infty} \frac{dk_0}{2\pi} \frac{m}{2iM^2} \left[\partial_k \langle \bar{\chi} \gamma^k \gamma^0 T^a \chi \rangle_M^A - i \mathcal{A}_{2n}^a \right] . \tag{33}$$

Here we have made the assumption that the gauge field commutes with T^a for a fixed value of a. The general result will be deduced via covariance arguments. Typically, the first term within the curly bracket of (33) is very complicated (non-local, non-analytical in A_μ), and cannot be evaluated in closed form for arbitrary fields. In the following we concentrate on the more interesting anomaly term. It is this term which is responsible for the induced vacuum charge

$$Q^a = \int d^{2n}x \, j^{0a}(x) ,$$

because, provided the field falls off sufficiently rapidly at infinity, the non-anomalous part of (33) can be eliminated using Stoke's theorem. Furthermore, this term vanishes for constant fields and, again, the anomaly is the only contribution. Finally, it can be shown that also for $m \to \infty$ only the second term survives. We shall see that the anomaly will lead to the Chern-Simons term in the action, whereas the ignored terms are those of normal parity. For n=1, the first (second) term of (33) corresponds to the first (second) term of (17).

In the sequel j^μ_a stands for the anomalous current alone. Doing a simple integration and imposing gauge and Lorentz covariance, Eq. (33) implies:

$$j^{\mu a}(x) = -\frac{1}{4} \frac{m}{|m|} K_n \, \epsilon^{\mu_1\mu_2\ldots\mu_{2n}} \, \mathrm{tr} \left(T^a F_{\mu_1\mu_2} \ldots F_{\mu_{2n-1}\mu_{2n}} \right) . \tag{34}$$

(For the singlet current the generator T^a has to be omitted.) This equation is the main result of our computation. It enables us, for instance, to calculate the vacuum charge of arbitrary classical field excitations. Specializing (34) for QED_3 , we find

$$j^\mu(x) = \frac{1}{8\pi} \frac{m}{|m|} \epsilon^{\mu\alpha\beta} F_{\alpha\beta}(x) .$$

Because of the ϵ-tensor, the spatial components of j^μ are determined by the electric field F_{0i} alone. Obviously, j^1 is proportional to E_2 , i.e. the current is perpendicular to the electric field. The vacuum shows a quantum Hall effect!

Introducing Lie-algebra-valued differential forms

$$A = iA_\mu\, dx^\mu = iA^a_\mu\, T^a\, dx^\mu$$

and

$$F = dA + A\wedge A = \tfrac{1}{2} F_{\mu\nu}\, dx^\mu\wedge dx^\nu = \tfrac{1}{2} F_{\mu\nu}{}^a\, T^a\, dx^\mu\wedge dx^\nu$$

we obtain the simple formula

$$Q^a_{2n+1} = -\frac{1}{2}\frac{m}{|m|}\left[\frac{i}{2\pi}\right]^n \frac{1}{n!}\int_{(2n)} \mathrm{tr}\,(T^a F^n) \tag{35}$$

and similarly for the singlet charge:

$$Q_{2n+1} = -\frac{1}{2}\frac{m}{|m|}\left[\frac{i}{2\pi}\right]^n \frac{1}{n!}\int_{(2n)} \mathrm{tr}\, F^n = -\frac{1}{2}\frac{m}{|m|}\int_{(2n)} \mathrm{tr}\,\exp\left[\frac{1}{2\pi} F\right]$$

$$= -\frac{1}{2}\frac{m}{|m|}\int_{(2n)} \mathrm{ch}\,(F)\ . \tag{36}$$

All integrals have been performed over 2n-dimensional x^k-space. The notation in the last two lines means that the exponential should be expanded in a power series, and that only the term containing the 2n-dimensional volume form has to be considered. We find that Q_{2n+1} is essentially given by the Chern character $\int \mathrm{ch}(F) = \int \exp(\frac{i}{2\pi} F)$. This means that it is a topological invariant[5], i.e. the charge does not change under local deformations of the background field. It is only the global (the topological) properties of A_μ which are relevant. It is well known[5] that $\int \mathrm{ch}(F)$ is precisely the index of the Dirac operator $\not{D}_{2n}$, i.e. the difference in the number of right-handed (n_+) and left-handed (n_-) zero modes (recall that chirality is defined with respect to γ^0).
Hence we have:

$$Q_{2n+1} = -\frac{1}{2}\frac{m}{|m|}\,\mathrm{index}\,[\not{D}_{2n}(A)] \equiv -\frac{1}{2}\frac{m}{|m|}\left(n_+[A] - n_-[A] \right). \tag{37}$$

It is now obvious that Q_{2n+1} is integer or half-integer. Eq. (37) suggests that the only spectral properties of $\not{D}_{2n}$ which have an impact on the charge are its zero modes. How this comes about can be nicely studied in our formalism. We start from the singlet version of (29), perform the k_o-integration, and integrate over all x^k, to arrive at

$$Q_{n+1} = -(4\pi)^{-\frac{1}{2}}\, m \left.\frac{d}{ds}\right|_o \frac{s}{\Gamma(s+1)} \int_0^\infty dt\, t^{s-\frac{1}{2}}\, e^{-m^2 t}\, \mathrm{Tr}\, [\gamma^o e^{-\not{D}_{2n}^2 t}] . \quad (38)$$

The symbol "Tr" denotes a summation over spinor and group indices as well as an integration over x^k. In this representation we recover a mathematical object which is well known from the general theory of index theorems[5,17], viz. the trace of the heat-kernel of an elliptic differential operator. Here we only give a simple argument why this trace is equal to the index of $\not{D}_{2n}$. To this end, let us consider a complete set of orthonormalized eigenfuctions of $\not{D}_{2n}$:

$$\not{D}_{2n}\Phi_m = \lambda_m\Phi_m \quad , \qquad \lambda_m \in \mathbb{R} \ .$$

The trace now reads

$$\mathrm{Tr}\, [\gamma^o e^{-\not{D}_{2n}^2 t}] = \sum_m \int d^{2n}x\, \Phi_m^+(x)\, \gamma^o\, e^{-\not{D}_{2n}^2 t}\, \Phi_m(x)$$

$$= \sum_m e^{-\lambda_m^2 t} \int d^{2n}x\, \Phi_m^+(x)\, \gamma^o\, \Phi_m(x) \ . \quad (39)$$

Because γ^o anticommutes with $\not{D}_{2n}$ we have

$$\not{D}_{2n}(\gamma^o\Phi_m) = -\lambda_m(\gamma^o\Phi_m) \ .$$

Since the euclidian operator $\not{D}_{2n}$ is hermitian with respect to the scalar product

$$(\Phi_1,\Phi_2) = \int d^{2n}x\, \Phi_1^+(x)\, \Phi_2(x) \ ,$$

this implies that Φ_m and $\gamma^o\Phi_m$ are orthogonal whenever λ_m is non-zero. Therefore (39) reduces to a sum over zero modes:

$$\mathrm{Tr}\,[\gamma^o e^{-\not{D}_{2n}^2 t}] = \sum_{\{\Phi_m | \not{D}_{2n}\Phi_m = 0\}} \int d^{2n}x\ \Phi_m^+(x)\gamma^o\Phi_m(x) \ . \tag{40}$$

Because $\not{D}_{2n}\Phi = 0$ implies $\not{D}_{2n}\frac{1}{2}(1\pm\gamma^o)\Phi = 0$, the zero modes can be assumed to have a definite chirality. Hence the sum (40) evaluates the difference between the number of positive and negative eigenvalues. We have thus shown that

$$\mathrm{Tr}\,[\gamma^o e^{-\not{D}_{2n}^2 t}] = \mathrm{index}\ \not{D}_{2n} \ . \tag{41}$$

Using this result it is trivial to perform the integral in (38) and to take the derivative at s=0. As expected, the answer is again Eq. (37).

To shed some light on its physical origin we represent the vacuum charge in still another way. Introducing the Hamiltonian

$$H = \gamma^o(-i\not{D}_{2n}+m) \ , \tag{42}$$

Eq. (38) reads

$$Q_{2n+1} = -\,(4\pi)^{-\frac{1}{2}}\ \frac{d}{ds}\bigg|_0 \frac{s}{\Gamma(s+1)} \int_0^\infty dt\ t^{s-\frac{1}{2}}\ \mathrm{Tr}\,[H\,e^{-H^2 t}] \ . \tag{43}$$

In terms of the eigenfunctions of the Hamiltonian $(H\psi_n = E_n\psi_n)$ this can be evaluated to yield

$$Q_{2n+1} = -\,\tfrac{1}{2} \lim_{s\to 0^+} \eta(H,s) \ . \tag{44}$$

Here we have introduced the η-invariant of Atiyah, Patodi and Singer[18], defined by

$$\eta(H,s) = \sum_n \mathrm{sign}(E_n)\ |E_n|^{-s} \ . \tag{45}$$

As in the case of the ζ-function, the series converges for sufficiently large values of Re(s) and can be analytically continued to a meromorphic function on the whole complex s-plane. It is a regularized version of the difference between the number of positive and negative eigenvalues of H. If we naively interchange the limit in (44) with the sum in (45), we have

$$Q_{2n+1} = -\frac{1}{2}\left[\sum_{E_n > 0} 1 - \sum_{E_n < 0} 1\right] . \tag{46}$$

In terms of the Hamiltonian it is the spectral asymmetry, i.e. the difference in the number of positive and negative eigenvalues, which determines the charge. The formal equation (46) can be used to calculate in a heuristic way the charge induced by a constant magnetic field in 3 dimensions[19]. We assume x^k-space to be a large square of side length L and set $F_{21} = -F_{12} = B = \text{const}$. With the γ^μ's represented by Pauli matrices the Hamiltonian reads

$$H = -i\sigma_2(\partial_1 - ieBx_2) + i\sigma_1\partial_2 + m\sigma_3 . \tag{47}$$

The eigenvalues of its normalizable (for $L \to \infty$) zero modes are

$$\begin{aligned} E(n,\pm) &= \pm\,[2neB + m^2]^{\frac{1}{2}} , \qquad n = 1,2,3,\ldots \\ E(0,+) &= +m . \end{aligned} \tag{48}$$

Their degeneracy is equal to $(eB/2\pi)L^2$. Note that there is no state with $E = -m$. Because the eigenvalues $E(n,\pm)$ are symmetric around $E = 0$, they do not contribute to (46). The spectral asymmetry is due to the unpaired $E(0,+)$ mode:

$$\begin{aligned} Q_3 &= -\frac{1}{2}\,\frac{eB}{2\pi}L^2 \cdot \begin{cases} +1 & \text{if } m > 0 \\ -1 & \text{if } m < 0 \end{cases} \\ &\equiv -\frac{1}{2}\,\frac{m}{|m|}\,\frac{eB}{2\pi}L^2 \\ &= -\frac{1}{2}\,\frac{m}{|m|}\,\frac{e}{4\pi}\int d^2x\, \epsilon_{ij}\, F^{ij} . \end{aligned} \tag{49}$$

This is the same as (36) for n=1. With this example we have now made contact with our discussion in the previous section. The level $E(0,+)$, which becomes a zero mode for $m \to 0$, corresponds to the mode Φ of the model defined by (1).

We note in passing that the above methods can be generalized by including a time-independant background gravitational field interacting with the fermions[16]. Then the metric can always be brought to the form

$$g_{\mu\nu}(x) = \begin{bmatrix} 1 & 0 \\ 0 & g_{ij}(\vec{x}) \end{bmatrix} .$$

For the induced vacuum charge one obtains

$$Q_{2n+1} = - \frac{1}{2} \frac{m}{|m|} \int_{\mathcal{M}_{2n}} ch(F) \wedge \hat{A}(\mathcal{M}_{2n}) .$$

The second factor in the integrand is the $\hat{A}$ - polynomial of the spatial manifold $\mathcal{M}_{2n}$ with metric $g_{ij}(\vec{x})$. In terms of the curvature 2-forms of g_{ij} , i.e.

$$\Omega \equiv (\Omega^i{}_j) \equiv (\tfrac{1}{2} R^i{}_{jkl} \, dx^k \wedge dx^l) ,$$

it is defined by the formal power series expansion of

$$\hat{A}(\mathcal{M}_{2n}) = \left[\det \frac{(\Omega/4\pi)}{\sinh(\Omega/4\pi)}\right]^{\frac{1}{2}} .$$

In particular, for $n=2$ and $A_\mu=0$, one finds

$$Q_5 = \frac{1}{2} \frac{m}{|m|} \frac{1}{768\pi^2} \int d^4x \, \epsilon^{ijkl} R_{nmij} R^{nm}{}_{kl} .$$

Examples of topologically non-trivial field configurations with $Q_5 \neq 0$ are the magnetic monopole solutions of the 5-dimensional Kaluza-Klein theory[16]. In this case one finds that the charge is quantized in units of 1/24.

THE EFFECTIVE ACTION

In this section we determine the part $\Gamma_{2n+1}[A]$ of the effective action, i.e. of the logarithm of the fermionic determinant, whose functional derivative is the anomalous vacuum current:

$$j^{\mu a}(x) = - \frac{\delta\Gamma_{2n+1}[A]}{\delta A^a_\mu(x)} \; ; \quad \Gamma_{2n+1}[0] = 0 \; . \tag{50}$$

For a given current the functional Γ_{2n+1} can be constructed in an elementary way (cf. Refs. [13] and [15]). The result is

$$\Gamma_{2n+1}[A] = \int d^{2n+1}x \; \mathcal{L}_{2n+1}(A(x)) \; ,$$

with the effective Lagrangian

$$\mathcal{L}_{2n+1}(A) = - \int_0^1 dt \; A^{\mu a}(x) \; j^a_\mu(tA(x)) \quad . \tag{51}$$

Inserting (34) for j^a_μ yields

$$\mathcal{L}_{2n+1}(A) = - \frac{m}{|m|} \frac{\pi}{n!} \left(\frac{i}{2\pi}\right)^{n+1} * \int_0^1 dt \; tr[A(tdA + t^2A^2)^n] \; . \tag{52}$$

(Here $*$ is the Hodge operator and $\int dt$ denotes an ordinary integration rather than a differential form.) The integral appearing in (52) is well known from the study of the algebraic properties of anomalies in even dimensional spaces[14]. In (2n+2)-dimensions the abelian anomaly, i.e. the anomaly of the singlet current, is essentially given by the Chern form

$$\Omega_{2n+2} = tr \; F^{n+1} \; .$$

This is a closed form. Therefore, locally it can be written as the exterior derivative of a (2n+1)-form, viz. the Chern-Simons form ω_{2n+1} :

$$\Omega_{2n+2} = d\omega_{2n+1} \; .$$

This form can be obtained from

$$\omega_{2n+1} = (n+1) \int_0^1 dt \; tr[A(tdA + t^2A^2)^n] \quad . \tag{53}$$

Comparison with (52) shows that Γ_{2n+1} is essentially given by the Chern-Simons term

$$\Gamma_{2n+1}[A] = -\frac{m}{|m|}\frac{\pi}{(n+1)!}\left[\frac{i}{2\pi}\right]^{n+1}\int \omega_{2n+1}(A) . \qquad (54)$$

This is the desired result. It shows that for all n the effective action belonging to the anomalous part of the vacuum current is given by the Chern-Simons term of the respective dimensionality. In particular, for n = 1, one has

$$\int\omega_3 = \int tr(A\wedge dA + \frac{2}{3} A\wedge A\wedge A) .$$

Expressing this in terms of the components A_μ^a, one finds (up to a normalization constant) nothing but the functional W[A] of the first section. This shows that gauge fields, which at the classical level do not have a topological mass term à la Deser et al.[1], acquire such a mass term via radiative corrections when they are coupled to fermions.

CONNECTIONS AMONG DIFFERENT ANOMALIES

Taken together with our knowledge about chiral anomalies in even dimensions the above results can be summarized as follows. In spaces of even dimension there exists a chiral anomaly which manifests itself as a $tr(*F^n)$- or a $tr(T^a*F^n)$- term in the divergence of the axial vector current. The reason is the impossibility to find a regularization scheme which not only respects the mandatory gauge invariance, but also the chiral invariance. In odd dimensions there is no chirality, and hence no chiral anomaly. However, there is an anomaly in the vacuum current itself (rather than in its divergence): it contains a parity violating piece. The reason is that no regularization scheme can be found which respects parity and gauge invariance at the same time. As we have seen, these two types of anomalies are not independant of each other. Let us start in D = 2n dimensions. Here we have a non-abelian chiral anomaly in the divergence of the non-singlet current, Cf. Eqs. (31) and (32). Eq. (33) shows that this anomaly generates the parity violating part of the vacuum current of the (2n+1)-dimensional theory; the associated effective Lagrangian is the Chern-Simons term ω_{2n+1}. However, the exterior differential of this form, i.e. $d\omega_{2n+1} = \Omega_{2n+2} \equiv \mathrm{tr}\, F^{n+1}$, is nothing but the singlet current anomaly ("abelian anomaly") in (2n+2)-dimensions. Schematically:

D=2n : non-abelian anomaly $\sim \mathrm{tr}(T^a * F^n)$

D=2n+1: vacuum current contains term $\sim \mathrm{tr}(T^a * F^n)$
associated effective Lagrangian is $\sim *\omega_{2n+1}$

D=2n+2: abelian anomaly $\sim *d\omega_{2n+1} = \mathrm{tr} * F^{n+1}$

Up to now these interrelations seem to have a purely mathematical character. In the next section we shall see that there are systems which exhibit a physical interplay between different types of anomalies. Similar connections between anomalies in different dimensions were found by Zumino et al.[14], who used differential geometric methods to solve the Wess-Zumino consistency conditions.

AN APPLICATION: COSMIC STRINGS AND DOMAIN WALLS

As was first discussed by Callan and Harvey[20], cosmic strings and domain walls[21] provide an interesting laboratory to study different kinds of anomalies within one physical system. In the first part of this section, we show how cosmic strings relate the abelian anomaly in (2n+2)-dimensions to the non-abelian anomaly in 2n dimensions. Then, in the second part, it is demonstrated that domain walls similarly connect the parity violating anomaly in three dimensions with the non-abelian chiral anomaly in two dimensions.

To keep the discussion simple, we first concentrate on cosmic strings in 4 dimensions. Following Ref. [20], we consider a theory containing a complex scalar field $\Phi = \Phi_1 + i\Phi_2$ with a non-zero vacuum expectation value v (which could arise from a quartic potential of the $\mu^2 < 0$ type). A string configuration of this field has the form

$$\Phi(x) = f(\rho)\, e^{in\varphi} , \qquad (55)$$

where the string is assumed to lie along the x^1-axis, and ρ and φ are polar coordinates in the orthogonal plane: $x^2 = \rho \cos\varphi$, $x^3 = \rho \sin\varphi$. The function f has to interpolate between $f(0) = 0$ and $f(\infty) = v$. Its precise form is unimportant for our purposes. Next we couple Φ to a Dirac spinor field ψ according to

$$\mathcal{L}_{int} = \bar{\psi}(\Phi_1 + i\gamma_5\Phi_2)\psi \, , \tag{56}$$

and study the solution of the Dirac equation

$$[i\partial\!\!\!/ - (\Phi_1 + i\gamma_5\Phi_2)]\psi = 0 \tag{57}$$

in the background of our string. Introducing

$$\partial\!\!\!/^{int} = \gamma_a\partial^a \quad (a=0,1) \, , \quad \gamma^{int} \equiv -\gamma^0\gamma^1 \, , \quad \gamma^{ext} \equiv -i\gamma^2\gamma^3 \tag{58}$$

so that

$$\gamma_5 = \gamma^{int}\gamma^{ext} \, , \tag{59}$$

as well as the chirality projections

$$\psi_\pm = \tfrac{1}{2}(1 \pm \gamma_5)\psi$$

the Dirac equation for axial symmetric ψ's reads (in the following we set $n=1$):

$$i\partial\!\!\!/^{int}\psi_- + i\gamma^2(\cos\varphi + i\gamma^{ext}\sin\varphi)\partial_\rho\psi_- - f(\rho)\, e^{i\varphi}\psi_+ = 0 \, ,$$

$$i\partial\!\!\!/^{int}\psi_+ + i\gamma^2(\cos\varphi + i\gamma^{ext}\sin\varphi)\partial_\rho\psi_+ - f(\rho)\, e^{-i\varphi}\psi_- = 0 \, . \tag{60}$$

The solutions are easily found:

$$\psi_-(x) = \eta(x^{int}) \exp\left[-\int_0^\rho f(\rho')\, d\rho'\right] , \quad \psi_+(x) = -i\gamma^2\psi_-(x) \, . \tag{61}$$

Here η is a spinor which depends on the "internal" coordinates x^a $(a=0,1)$ only, and fulfills

$$i\partial\!\!\!/^{int}\eta = 0 \, , \tag{62}$$

together with

$$\gamma^{int}\eta = -\eta \,. \tag{63}$$

Because γ^{int} is the analogue of γ_5 for an observer living in a two-dimensional (0,1) world on the string, this observer would interpret the solutions (61) as massless particles of definite chirality running along the string. (This "dimensional reduction" is formally very similar to the treatment of fermions in Kaluza-Klein theory[22].)

If we couple the (Dirac) fermion ψ (charge e) and the scalar Φ (charge 2e) to an abelian gauge field A_μ, this results in an anomaly-free theory, i.e. we have a conservation law $\partial_\mu j^\mu = 0$ for the gauge charge. On the other hand, the string zero modes are chiral two-dimensional spinors; applying an external gauge field, their current fails to be conserved:

$$\partial^a J_a(x^0,x^1) = \frac{e}{2\pi}\,\epsilon^{ab}\partial_a A_b(x^0,x^1;x^2=x^3=0) \,, \qquad (a,b=0,1) \,. \tag{64}$$

(The fact that there is an unbalance between $\gamma^{int}= +1$ and $\gamma^{int}= -1$ is guaranteed by the index theorem.) Hence the charge of the massless modes running along the string is not conserved. Assuming a purely electric, constant field along the x^1-axis, say, the charge on the string (per unit length) changes as

$$\dot{Q}= \frac{eE}{2\pi} \,. \tag{65}$$

Because the complete theory is anomaly free, there must be a possibility for the string to exchange charge with the outside world. On the other hand, it is hard to see how this can happen, because, due to the Yukawa coupling (56), outside the string the fermions have a mass of order v, which can be made arbitrarily large. Formulated at the level of the effective action, we know that the zero modes spoil the gauge invariance of the fermionic determinant (ω is an infinitesimal parameter):

$$\delta_\omega(\ln \det iD\,)_{ZM} = \frac{ie}{2\pi}\int d^2x\,\epsilon^{ab}\partial_a A_b \,. \tag{66}$$

The problem is to find other contributions to $\det(i\not{D})$ which cancel (66) and restore charge conservation. In the presence of an external gravitational field, a similar discussion applies to the conservation of the energy-momentuum tensor. The solution to this puzzle is that the external A_μ-field induces a current in the vacuum of the quantized scalar field. For slowly varying fields it is given by

$$\langle J_\mu \rangle = -i \frac{e}{16\pi^2} \epsilon_{\mu\nu\alpha\beta} \frac{\Phi^* \partial^\nu \Phi - \Phi \partial^\nu \Phi^*}{|\Phi|^2} F^{\alpha\beta} , \tag{67}$$

in particular, for $\Phi(x) = v \exp(i\theta(x))$, one finds

$$\langle J_\mu \rangle = \frac{e}{8\pi^2} \epsilon_{\mu\nu\alpha\beta} (\partial^\nu \theta) F^{\alpha\beta} . \tag{68}$$

For the string, outside its core, we have $\Phi = v \exp(i\varphi)$. In the case of an electric field in the x^1-direction, this yields

$$\langle J_\rho \rangle = \frac{eE}{2\pi} \cdot \frac{1}{2\pi\rho} \tag{69}$$

for the radial component, and zero for the other components. Note that this current is perpendicular to the electric field. Here again we find a quantum Hall effect, this time with bosonic charge carriers (Cf. the remarks following eq. (34)). Integrating (69) over a cylinder surrounding the string, we see that the charge transported onto the string (per unit time and unit length) is precisely equal to (65). The "Hall" current (67), which is essentially a 3-dimensional quantity because it has no x^1-component, is exactly what is needed to compensate for the charge non-conservation due to the chiral anomaly. The current (68) can be derived from the action

$$S_{eff} = - \frac{e^2}{16\pi^2} \int d^4x \, \epsilon_{\mu\nu\alpha\beta} \, \partial^\mu \theta \, A^\nu \, F^{\alpha\beta} \tag{70a}$$

$$" = " \quad \frac{e^2}{32\pi^2} \int d^4x \, \theta(x) \, \epsilon_{\mu\nu\alpha\beta} \, F^{\mu\nu} \, F^{\alpha\beta} . \tag{70b}$$

The second version of (70) shows that $\langle J_\mu \rangle$ is related to the chiral anomaly in 4 dimensions. But strictly speaking, (70b) is meaningless, because θ is a phase angle and therefore only defined modulo 2π. Evaluating the change of (70a) under gauge transformations, it turns out that it is the negative of the contribution (66) from the chiral zero modes. This demonstrates once more that the inclusion of the Φ-vacuum current restores gauge invariance.

It is easy to generalize the discussion to a non-abelian gauge field. The effective action for $\langle J_\mu \rangle$ then reads

$$S_{eff} = -\frac{1}{8\pi^2}\int d^4x\,(\partial^\mu\theta)\cdot\epsilon_{\mu\nu\alpha\beta}\,\mathrm{tr}\,(\,A^\nu\partial^\alpha A^\beta + \frac{2}{3}A^\nu A^\alpha A^\beta\,) \tag{71a}$$

$$\text{"="} \quad \frac{1}{32\pi^2}\int d^4x\,\theta(x)\,\epsilon_{\mu\nu\alpha\beta}\,\mathrm{tr}\,(\,F^{\mu\nu}F^{\alpha\beta}\,)\,. \tag{71b}$$

In the first version we recognize the Chern-Simons form, in the second the abelian anomaly. For the gauge variation of S_{eff} one finds

$$\delta_\omega S_{eff} = \frac{1}{2\pi}\int d^2x\,\mathrm{tr}\,(\,\omega\,\epsilon^{ab}\partial_a A_b\,)\,. \tag{72}$$

This is just what is needed to compensate for the non-abelian anomaly of the two-dimensional zero modes on the string. The analysis can be extended to strings living in (2n+2)-dimensions. Then there are 2n internal coordinates x^a for the "world" on the string. It turns out that the coupling between θ and the gauge field is always given by the abelian chiral anomaly in (2n+2)-dimensions, whereas the charge non-conservation on the string is dictated by the abelian anomaly in 2n dimensions.

In the above example the physical connection between the 2- and 4-dimensional chiral anomalies was provided by a bosonic Hall current, which is intrinsically 3-dimensional (it lives in the plane perpendicular to the string). We now turn to a system which features the parity violating current investigated earlier together with a 2-dimensional chiral anomaly: fermions in the background field of a domain wall[20,21].

We consider the theory defined by

$$\mathcal{L} = \bar\Psi(i\partial\!\!\!/ - \kappa\Phi)\psi + \frac{1}{2}\,\partial_\mu\Phi\,\partial^\mu\Phi - V(\Phi) \tag{73}$$

in three dimensions. The field Φ is a real pseudo-scalar. For $V(\Phi) = V(-\Phi)$ the Lagrangian is invariant under parity transformations (recall that $\bar\Psi\psi$ is odd under P). We assume that V is of the symmetry-breaking type; hence there exist two vacua with $\langle\Phi\rangle = v$ and $\langle\Phi\rangle = -v$ ($v>0$), respectively. Domain wall solutions

$$\Phi(x^0,x^1,x^2) = \Phi_0(x^2) \tag{74}$$

are similar to the familiar kinks in 2 dimensions. The function Φ_0 is required to interpolate between the two vacua when going from $x^2 = -\infty$ to $x^2 = +\infty$, i.e. we have $\Phi_0(-\infty) = -v$ and $\Phi_0(+\infty) = +v$. This is the only information about Φ_0 as a background. Introducing

$$\partial\!\!\!/^{int} = \gamma^a \partial_a \quad (a = 0,1), \qquad \gamma^{int} = -\gamma^0\gamma^1, \qquad \gamma^2 = i\gamma^{int} \tag{75}$$

the equation of motion reads

$$[i\partial\!\!\!/^{int} - \gamma^{int}\partial_2 - \kappa\Phi_0(x_2)]\,\psi(x) = 0\ . \tag{76}$$

Its solution is

$$\psi(x) = \eta(x^{int})\ \exp\left[-\int_0^{x^2} d\tilde{x}^2\ \Phi_0(\tilde{x}^2)\right], \tag{77}$$

where

$$i\partial\!\!\!/^{int}\eta(x^{int}) = 0\ , \qquad \gamma^{int}\eta = \eta\ . \tag{78}$$

The coordinates x^{int} refer to the two-dimensional (0,1)-world on the wall. (In our example the wall has only one spacelike direction.) Similar to the case of the cosmic string, we find massless chiral excitations, which are localized in the neighborhood of the wall. If we now couple the fermions to a (dynamical) Yang-Mills field we know that at least two things will happen. First, via the $d=2$ chiral anomaly the zero modes make the fermion determinant non-gauge-invariant:

$$\delta_\omega(\ln\det iD\!\!\!/\,)_{ZM} = \frac{1}{4\pi}\int d^2x\ \epsilon^{ab}\ \mathrm{tr}\ (\ \omega\ F_{ab}(x^2{=}0)\ ), \quad (a,b = 0,1). \tag{79}$$

Second, a mass term for the gauge bosons is dynamically generated. For slowly varying Φ_0 it is obtained from (54) for $n=1$ by replacing m by $\Phi_0(x^2)$:

$$\Gamma_3 = \frac{1}{16\pi}\int d^3x\ \frac{\Phi_0(x^2)}{|\Phi_0(x^2)|}\ \epsilon^{\mu\nu\rho}\ \mathrm{tr}\ (\ A_\mu\partial_\nu A_\rho + \frac{2}{3}\,iA_\mu A_\nu A_\rho\ )\ . \tag{80}$$

Because the complete theory is anomaly free, (79) must be canceled by other contributions to the effective action. It turns out that it is precisely the gauge-boson mass-term (80) which provides these contributions. Evaluating its gauge variation one obtains the negative of

(79). Thus we can say that the 2-dimensional (non-abelian) chiral anomaly and the 3-dimensional parity violating anomaly cannot exist without each other in this model. For the complete theory to be consistent, either both of them have to be present, or both absent. This can be seen nicely at the level of the currents. Taking, for example, a purely electric, abelian external field along the x^1-axis, the anomaly (79) implies that the charge per unit length on the wall is not conserved, but changes according to

$$\dot{Q} = \frac{eE}{2\pi} . \tag{81}$$

On the other hand, the E-field induces a vacuum current which can be obtained from the equation following (34):

$$j^{\mu}(x) = \frac{e}{8\pi} \frac{\Phi_0(x^2)}{|\Phi_0(x^2)|} \epsilon^{\mu\alpha\beta} F_{\alpha\beta}(x) . \tag{82}$$

Approximating $\Phi_0(x^2)/|\Phi_0(x^2)|$ by the step function $\epsilon(x^2)$, we get for our example

$$j^2 = -\frac{1}{2} \epsilon(x^2) \frac{eE}{2\pi} \tag{83}$$

and $j^0=j^1=0$. Integrating this current to obtain the charge transported onto the wall, or away from it, per unit length and unit time, we find nothing but (81)! This shows again that the two types of anomalies, chiral and parity violating, imply each other and cannot exist separately. This gives a physical meaning to the mathematical structures described in the previous section.

For a further discussion of these models we refer to Callan and Harvey[20]. A numerical investigation of similar systems using Wilson lattice fermions is presented in Ref. [23]. A realization of the parity violating anomaly in condensed matter physics is described in Ref. [24].

ACKOWLEDGEMENT

I would like to thank J. Magpantay for interesting discussions and for a critical reading of the manuscript.

REFERENCES

1. S. Deser, R. Jackiw, S. Templeton, Phys. Rev. Lett. 48: 975 (1982); Ann. Phys. 140: 372 (1982).
2. R. Jackiw, in: "Asymptotic Realms of Physics", A. Guth, K. Wan, R. Jaffe, eds., MIT-Press (1983).
3. R. Jackiw, Comm. Nucl. Part. Phys. 13: 15 (1984).
4. R. Jackiw, J. R. Schrieffer, Nucl. Phys. B190: 253 (1981).
5. A good introduction is T. Eguchi, P. Gilkey, A. Hanson, Phys. Rep. 66: 213 (1980).
6. See, for example, A. Actor, Rev. Mod Phys. 51: 461 (1979).
7. R. Jackiw, in: "Relativity, Groups and Topology II", B. S. DeWitt, R. Stora, eds., North-Holland (1984).
8. R.Jackiw, Comm. Nucl. Part. Phys. 13: 141 (1984).
9. A. N. Redlich, Phys. Rev. Lett. 52: 18 (1984); Phys. Rev. D29: 2366 (1984).
10. A. J. Niemi, G. W. Semenoff, Phys. Rev. Lett. 51: 2077 (1983).
11. J.S. Dowker, R. Critchley, Phys. Rev. D13: 3224 (1976); S.W. Hawking, Comm. Math. Phys. 55: 133 (1977).
12. M. Reuter, Phys. Rev. D31: 1374 (1985).
13. M. Reuter, W. Dittrich, Phys. Rev. D33: 601 (1986).
14. Cf. the lectures in "Current Algebra and Anomalies", S.B. Treiman, R. Jackiw, B. Zumino, E. Witten, eds., World Scientific (1985) and the references therein.
15. W. Dittrich, M. Reuter, "Selected Topics in Gauge Theories", Lecture Notes in Physics, Vol. 224, Springer (1986).
16. M.B. Paranjape, G.W. Semenoff, Phys. Rev. D31: 1324 (1985).
17. P.B. Gilkey, "Invariance Theory, The Heat Equation and the Atiyah-Singer Index Theorem", Publish or Perish (1984).
18. M. Atiyah, V. Patodi, I. Singer, Bull. London Math. Soc. 5: 229, (1973).
19. R.J. Hughes, Phys. Lett. 148B: 215 (1984).
20. C.G.Callan, J.A. Harvey, Nucl. Phys. B250: 427 (1985).

21. A. Vilenkin, Phys. Rep. 21: 263 (1985).

22. E. Witten, in: "Proc. of the 1983 Shelter Island Conference", R. Jackiw et al.,eds., MIT Press (1985).

23. E. Dagotto, Phys. Rev. D34: 2457 (1986).

24. E. Fradkin, E, Dagotto, D. Bojanowsky, Phys. Rev. Lett. 57: 2967 (1986).

III. SUPERTHEORIES

SUPERSYMMETRY IN
TWO SPACE-TIME DIMENSIONS

M. F. Sohnius

SUPERGRAVITY AS A GAUGE THEORY

A. C. Hirshfeld

STRINGS AND SUPERSTRINGS

H. Nicolai

SUPERSYMMETRY IN TWO SPACE-TIME DIMENSIONS

Martin F. Sohnius

Mathematics Department
King's College, University of London
London WC2R 2LS

ABSTRACT

Motivated by superstrings, supersymmetry in one space- and one time-dimension is developed. In this special situation supersymmetry in the right-moving sector need not be accompanied by supersymmetry in the left-moving sector. This is crucuial for the theory of the heterotic string. Both (1,0)- and (1,1)-supersymmetry are discussed. Finally, the Lagrangian for the supersymmetric σ-model is given, both in the superspace and the component formalism, and its reparametrisation invariance is examined.

1 INTRODUCTION

These lecture notes cover the second half of my lectures at the School. The first half was concerned with properties of spinors in space-times with arbitrary dimensions and arbitrary signature of the metric, and a compact set of notes on this subject can be found in the appendix of my *Physics Report* "Introducing Supersymmetry" (Physics Reports 128: 39 (1985)).

Traditionally, "physics" is thought to take place in three dimensions of space and one of time, and all realistic model building in theoretical physics used to be confined to that case. In the case of supersymmetric theories, it was indeed the successful generalisation of supersymmetry from two to four space-time dimensions by J. Wess and B. Zumino in 1973 which started the bandwagon rolling that now carries a good proportion of the world's community of theoretical high energy physicists.

Almost fifteen years of research, first into supersymmetry, then into supergravity, together with the resurrection of Kaluza-Klein ideas (gravity in more than four dimensions as source of all interactions), have both created an enormous new wealth of theoretical tools and overcome some of the fears of dealing with "exotic" theories. A few years ago, the time was ripe for a re-examination of the almost forgotten string theories of the late 1960's. The ensuing discovery of supersymmetric string theories by M. Green and J. Schwarz then brought the topic of supersymmetry around full circle, albeit into a different Riemann sheet: the "old" supergauge symmetry in the two dimensions of the world sheet of certain fermionic strings had, after all, been the starting point of Wess' and Zumino's work. The superstring encompasses both that old supergauge theory and the "new" supersymmetry in higher dimensions. It is not believed that supersymmetry is manifest in the narrow window of phenomenology which looks like, and is perceived as, four dimensional space-time.

In the teaching of supersymmetry, it is a blessing that one now has the string theories as a good excuse for doing it in two dimensions: technical complications grow roughly like the number of spinor components and thus exponentially with the dimension. All the principal ideas, however, stay the same, and so one may as well teach the basics in two instead of four dimensions. This is why I am attempting here to give an introduction to supersymmetry in two dimensions.

The lecture notes are organised as follows: in section 2, the (n,m) superalgebras are given for two dimensional Minkowski space. Section 3 is devoted to representations, both in terms of multipets and of superfields, of the simplest of these algebras, that of (1,0) supersymmetry. Section 4 gives actions for two of these multiplets, again in superspace as well as in components. It is found that supersymmetry in the right-moving sector need not be accompanied by supersymmetry in the left-moving one. This is special to two dimensions and crucial for the theory of the heterotic string. Section 5 is devoted to the larger (1,1) supersymmetry which more closely resembles the case of four and more dimensions. That section culminates in the Lagrangian for the (1,1) supersymmetric σ-model. Section 6, finally, contains an examination of properties of this σ-model: its reparametrisation invariance is shown as obvious in the superspace formulation, and then the lengthy derivation of the component version of the Lagrangian is carried through in detail. The lesson to be learned from this will be that superspace notation in two dimensions is indeed a powerful tool, much more so, actually, than in higher dimensions. To appreciate its power, the reader is invited to compare the Lagrangians of Eqs. (6.28) and (6.4).

2. THE SUPERALGEBRAS IN TWO DIMENSIONS

In space-times which allow *Majorana spinor charges* Q_α with

$$Q = C\overline{Q}^T = DQ^+ , \tag{2.1}$$

the usual *supersymmetry algebra* is

$$\{Q,\overline{Q}\} = 2\gamma^\mu P_\mu ; \quad [Q,P_\mu] = 0 ; \quad [P_\mu, P_\nu] = 0 . \tag{2.2}$$

The operators P_μ are the components of the usual relativistic energy-momentum vector.

For two dimensional Minkowski space, the difference between the number of time and space dimensions is zero and thus divisible by eight. Hence Majorana-Weyl spinors (chiral Majorana spinors) are a possibility. The dimension of a "generic" spinor in n-dimensional space-time is $2^{n/2}$ complex components, i.e. 2 for n=2. A Majorana-Weyl spinor thus has a single, real component. In detail, these properties become apparent if we now turn to an explicit representation of the Dirac matrices for this case.

With $\eta_{\mu\nu} \equiv \mathrm{diag}(1,-1)$, the Dirac-algebra

$$\{\gamma_\mu, \gamma_\nu\} = 2\eta_{\mu\nu} \tag{2.3}$$

is represented by the imaginary 2×2 matrices

$$\gamma_0 \equiv \begin{bmatrix} 0 & -i \\ i & 0 \end{bmatrix} ; \quad \gamma_1 \equiv \begin{bmatrix} 0 & i \\ i & 0 \end{bmatrix} . \tag{2,4}$$

The other relevant matrices are

$$\gamma_3 \equiv \gamma_0\gamma_1 = \begin{bmatrix} 1 & 0 \\ 0 & -1 \end{bmatrix}; \quad A = \gamma_0 = \begin{bmatrix} 0 & -i \\ i & 0 \end{bmatrix}; \quad C = -\gamma_0 = \begin{bmatrix} 0 & i \\ -i & 0 \end{bmatrix}; \quad D = CA^T = 1 \tag{2.5}$$

and we see that the Majorana condition (2.1) becomes a reality condition $Q = Q^+$, and the Weyl condition

$$Q = \tfrac{1}{2}(1+\gamma_3)Q \tag{2.6}$$

means that Q has only one component.

The *Lorentz Group* of our space-time has only one independent generator

$$M_{\mu\nu} = \epsilon_{\mu\nu} M , \quad \text{with } \epsilon_{\mu\nu} = -\epsilon_{\nu\mu} \text{ and } \epsilon_{01} = +1 . \tag{2.7}$$

From the general transformation law for a spinor Q,

$$[Q, M_{\mu\nu}] = \tfrac{1}{2}\sigma_{\mu\nu} Q \quad \text{with} \quad \sigma_{\mu\nu} \equiv \tfrac{i}{2}[\gamma_\mu, \gamma_\nu] , \tag{2.8}$$

we can calculate that its components

$$Q \equiv \begin{bmatrix} Q_+ \\ Q_- \end{bmatrix} \tag{2.9}$$

transform among themselves as follows:

$$[Q_\pm, M] = \pm \tfrac{i}{2} Q_\pm . \tag{2.10}$$

From the usual transformation law for the energy and momentum,

$$[P_\mu, M_{\rho\sigma}] = i\eta_{\mu\rho} P_\sigma - i\eta_{\mu\sigma} P_\rho , \tag{2.11}$$

we find that

$$[P_0 \pm P_1, M] = \pm\, i(P_0 \pm P_1) . \tag{2.12}$$

Since the combinations $P_0 \pm P_1$ transform similarly to $Q_\pm$, but with twice the weight, we choose the popular notation

$$P_{++} \equiv P_0 + P_1 ; \quad P_{--} \equiv P_0 - P_1 . \tag{2.13}$$

In two dimensions, the vector is not an irreducible representation of the rotation group: as the dual of a vector is again a vector (albeit an axial one), the self-dual and anti-self-dual components are representations in their own right. The light-cone components P_{++} and P_{--} are just these irreducible representations.

The anticommutator of Eqs. (2.2) becomes in this notation

$$Q_+Q_+ = P_{++} \; ; \quad Q_-Q_- = P_{--} \tag{2.14a}$$

$$\{Q_+,Q_-\} = 0 \; . \tag{2.14b}$$

We see that the "right-moving" and "left-moving" supersymmetries (Q_+ and Q_-) are completely independent of each other, due to Eq. (2.14b).

It is possible to have more than one supersymmetry of each type: if we have n right moving Q_+^i (i=1,.,n) and m left moving Q_-^I (I=1,.,m), then the algebra becomes

$$Q_+^i \, Q_+^j = \delta^{ij} \, P_{++} \; ; \quad Q_-^I \, Q_-^J = \delta^{IJ} P_{--} \tag{2.15a}$$

$$\{Q_+^i, Q_-^J\} = 0 \; , \tag{2.15b}$$

and is said to be that of (n,m)-extended supersymmetry.

One may very well have $n \neq m$. This situation, which reflects the complete independence of right and left moving sectors, is peculiar to two dimensions: it originates in the absence of a rotation which connects the right and left branches of the light-cone.

3. MULTIPLETS AND SUPERFIELDS OF (1,0) SUPERSYMMETRY

We now try to represent the algebra of (1,0) supersymmetry on quantum fields over two-dimensional space. The generators whose actions must be represented are P_{++} and P_{--}, one Q_+ but no Q_-. We replace the coordinates x^μ by light-cone coordinates

$$x^{++} \equiv x^0 + x^1 \; ; \quad x^{--} = x^0 - x^1 \tag{3.1}$$

and then have

$$\partial_{++} \equiv \frac{\partial}{\partial x^{++}} = \frac{\partial}{\partial x^0} + \frac{\partial}{\partial x^1} = \partial_0 + \partial_1 \; ; \quad \partial_{--} = \partial_0 - \partial_1 \tag{3.2}$$

so that for any quantum-field $\phi(x)$ the momenta $P_{\substack{++\\--}}$ generate translations along the $x^{\substack{++\\--}}$ directions:

$$[\phi(x), P_{\substack{++\\--}}] = i\partial_{\substack{++\\--}}\phi(x) \; . \tag{3.3}$$

We now derive the so-called *scalar multiplet*. We start with a real scalar field A(x) (scalar: $[A(x),M] = 0$). Then $-i[A(x),Q_+]$ must be a new real field which transforms like the (+) component of a spinor:

$$[A(x),Q_+] = i\psi_+(x) . \tag{3.4}$$

Further, the *anti*-commutator of ψ_+ with Q_+ is another real field:

$$\{\psi_+(x),Q_+\} = X_{++}(x) .$$

But from $Q_+Q_+ = P_{++}$ we get

$$i\partial_{++}A = [A,P_{++}] = [A,Q_+Q_+] = \{[A,Q_+],Q_+\} = i\{\psi_+,Q_+\} = iX_{++}$$

so that

$$\{\psi_+(x),Q_+\} = \partial_{++}A(x) . \tag{3.5}$$

Finally, we check that

$$[\psi_+,Q_+Q_+] = [\{\psi_+,Q_+\},Q_+] = [\partial_{++}A,Q_+] = i\partial_{++}\psi_+ \tag{3.6}$$

as required by the algebra and Eq. (3.3). We have thus found our first supermultiplet, consisting of fields A and ψ_+ with transformation laws (3.4)-(3.5).

Similarly, we can start with a spinor field $\lambda_\pm(x)$ of either chirality and derive a *spinor multiplet* from the equivalent of Eq. (3.4):

$$\{\lambda_\pm,Q_+\} = F_{\pm+} . \tag{3.7}$$

The general transformation law for $F_{\pm+}$,

$$[F_{\pm+},Q_+] = i\psi_{\pm++} ,$$

is again restricted by the algebra,

$$i\partial_{++}\lambda_\pm = [\lambda_\pm,P_{++}] = [\lambda_\pm,Q_+Q_+] = [\{\lambda_\pm,Q_+\},Q_+] = [F_{\pm+},Q_+] = i\ \psi_{\pm++}$$

so that

$$[F_{\pm+},Q_+] = i\partial_{++}\lambda_\pm . \tag{3.8}$$

Again, we can check that

$$[F_{\pm+},Q_+Q_+] = \{Q_+,[F_{\pm+},Q_+]\} = \{Q_+,i\partial_{++}\lambda_\pm\} = i\partial_{++}F_{\pm+} , \tag{3.9}$$

as required.

We note that the only differences between the scalar and spinor multiplets are an interchange in the roles of commutators and anticommutators and a few factors of i.

We could, of course, slam more Lorentz indices onto our fields: if their number is even, we have tensor multiplets with transformation laws like Eqs. (3.4)-(3.5), if it is odd, we have spinor multiplets with transformation laws like Eqs. (3.7)-(3.8). Note the additional index on Ψ and F: it is always a (+) and comes from the index on Q_+.

An elegant notation for all of this is the *superfield*. Let us introduce a real, constant spinor component θ and assume it to *anticommute* with all spinorial objects around (we don't bother here with mathematical existence theorems for such objects). Since, in particular, it anticommutes with itself, we have

$$\theta\theta = \tfrac{1}{2}\{\theta,\theta\} = 0 \tag{3.10}$$

and the most general power series in θ is

$$f(\theta) = f_0 + \theta f_1 \ . \tag{3.11}$$

We can introduce a derivative $\partial/\partial\theta$, defined by

$$\frac{\partial}{\partial\theta}(f_0 + \theta f_1) \equiv f_1 \ . \tag{3.12}$$

We now define a *scalar superfield* from our scalar multiplet by

$$\Phi(x,\theta_-) \equiv A(x) + i\theta_-\Psi_+(x) \ , \tag{3.13}$$

with θ_- such a real, constant, anticommuting variable. We then have, as for ordinary fields,

$$[\Phi,P_{\underline{++}}] = i\ \partial_{\underline{++}}\Phi \ , \tag{3.14}$$

but also a new commutator of Φ with Q_+ (note how in the following the commutator changes into an anticommutator when θ_- is pulled to the outside - this is because θ_- anticommutes with Q_+)

$$\begin{aligned}[\Phi,Q_+] &= [A,Q_+] + i\theta_-\{\Psi_+,Q_+\} = i\Psi_+ + i\theta_-\partial_{++}A \\ &= (\frac{\partial}{\partial\theta_-} + i\theta_-\partial_{++})(A + i\theta_-\Psi_+) \equiv r(Q_+)\Phi \ .\end{aligned} \tag{3.15}$$

The square of the differential operator $r(Q_+)$,

$$r(Q_+)\ r(Q_+) = \frac{\partial}{\partial\theta_-}\frac{\partial}{\partial\theta_-} + i\theta_-\partial_{++}\frac{\partial}{\partial\theta_-} + \frac{\partial}{\partial\theta_-}(i\theta_-\partial_{++}) - (\theta_-\partial_{++})^2 \ ,$$

simplifies considerably because the first and the last term vanish due to Eq. (3.10). Further, since for each θ which satisfies Eq. (3.10) we get

$$\frac{\partial}{\partial\theta}\theta = (\frac{\partial}{\partial\theta}\theta) - \theta\frac{\partial}{\partial\theta} = 1 - \theta\frac{\partial}{\partial\theta} \ , \tag{3.16}$$

we have

$$r(Q_+)\ r(Q_+) = i\partial_{++} \ , \tag{3.17}$$

so that $r(Q_+)$ and $r(P_{\underline{\underline{++}}}) = i\partial_{\underline{\underline{++}}}$ are a representation of the superalgebra. As required, we also have

$$[r(Q_+), r(P_{\underline{\underline{++}}})] = 0 \tag{3.18}$$

since $r(Q_+)$ contains no explicit $x^{\underline{\underline{++}}}$.

In general, we *define* a (*bosonic*) *superfield* to be a formal power series in θ_- which transforms under supersymmetry transformations as in Eq. (3.15).

The beauty of the superfield approach becomes apparent if we notice that, because $r(Q_+)$ is a *linear* differential operator, it observes the Leibniz rule and thus, e.g.,

$$[\Phi^2, Q_+] = \Phi[\Phi, Q_+] + [\Phi, Q_+]\Phi = 2\Phi r(Q_+)\Phi = r(Q_+)\Phi^2 \ ,$$

so that, more generally, any product of superfields is again a superfield.

Note that not every power series in θ_- is a superfield. E.g., if we look at $f \equiv i\theta_-\Psi_+$, we have $[f, Q_+] = i\theta_-\partial_{++}A$, but $r(Q_+)f = i\Psi_+$ which is by no means the same.

Everything said so far about superfields remains true if we attach an even number of Lorentz indices to $\Phi(x,\theta)$, i.e., to both $A(x)$ and $\Psi_+(x)$. But how about superfields of a spinorial nature with an odd number of additional spinor indices? We already have a spinor multiplet, so we try the ansatz

$$\Lambda_\cdot(x,\theta) \equiv \lambda_\cdot(x) \pm \theta_- \, F_{\cdot +}(x) \ ,$$

where the $\cdot$ stands for an odd number of spinor indices. We leave the sign between the two terms undetermined, and get from Eqs. (3.7)-(3.8)

$$\{\Lambda_\cdot, Q_+\} = F_{\cdot +} \pm i\theta_-\partial_{++}\lambda_\cdot = \pm\, r(Q_+)\Lambda_\cdot \ .$$

So far, everything is nicely consistent with the algebra for either choice of sign since

$$[\Lambda_\cdot, Q_+Q_+] = [\{\Lambda_\cdot, Q_+\}, Q_+] = \pm\, [r(Q_+)\Lambda_\cdot, Q_+]$$

$$= \pm\, r(Q_+)\{\Lambda_\cdot, Q_+\} = r^2(Q_+)\Lambda_\cdot = i\partial_{++}\Lambda_\cdot \ ,$$

but if we consider the product $\Lambda_\cdot\Phi$, we get

$$\{\Lambda_\cdot\Phi, Q_+\} = \Lambda_\cdot[\Phi, Q_+] + \{\Lambda_\cdot, Q_+\}\Phi = \Lambda_\cdot r(Q_+)\Phi \pm (r(Q_+)\Lambda_\cdot)\Phi \ .$$

Taking into account the minus-sign in the chain rule

$$r(Q_+)(\Lambda.\phi) = (r(Q_+)\Lambda.)\phi - \Lambda.r(Q_+)\phi \ , \tag{3.19}$$

this makes $\Lambda.\phi$ a decent superfield only for the choice of the minus-sign, giving us

$$\Lambda.(x,\theta) \equiv \lambda.(x) - \theta_- F_{.+}(x) \tag{3.20}$$

$$\{\Lambda.,Q_+\} = -r(Q_+)\Lambda. \quad . \tag{3.21}$$

A *fermionic superfield* is defined to transform like this.

To conclude this section, we look at the *derivatives* of superfields. If ϕ or Λ are superfields, then so are $\partial_{\underset{--}{++}}\phi$ and $\partial_{\underset{--}{++}}\Lambda.$, since, e.g.

$$[\partial_{++}\phi,Q_+] = \partial_{++}[\phi,Q_+] = \partial_{++}(r(Q_+)\phi) = r(Q_+)(\partial_{++}\phi) \ ,$$

as required. This works because $r(Q_+)$ commutes with $\partial_{\underset{--}{++}}$.

But can we find a spinorial derivative D, built using θ's, such that $D\phi$ is a fermionic superfield and $D\Lambda.$ a bosonic one? For this we require

$$- r(Q)(D\phi) = \{D\phi,Q\} = D[\phi,Q] = D(r(Q)\phi)$$

and

$$r(Q)(D\Lambda.) = [D\Lambda.,Q] = D\{\Lambda.,Q\} = - D(r(Q)\Lambda.)$$

i.e. in both cases that

$$\{D,r(Q)\} = 0 \ . \tag{3.22}$$

Since $\{\frac{\partial}{\partial\theta_-},r(Q_+)\} = i\partial_{++}$ and $\{\theta_-\partial_{++},r(Q_+)\} = \partial_{++}$ an answer is

$$D_+ = - i \frac{\partial}{\partial\theta_-} - \theta_-\partial_{++} \ . \tag{3.23}$$

The components of the superfields $D_+\phi$ and $D_+\Lambda.$ are

$$D_+\phi = \Psi_+ - \theta_-\partial_{++}A \tag{3.24}$$

$$D_+\Lambda. = iF._+ - \theta_- \partial_{++}\lambda. \quad . \tag{3.25}$$

Finally, we have

$$D_+D_+ = i\partial_{++} \ . \tag{3.26}$$

Our rules allow us to freely multiply and differentiate superfields, using the derivatives D_+, ∂_{++} and ∂_{--} , provided we carefully watch minus-signs when interchanging spinorial quantities.

4. INVARIANT ACTIONS

Our next task will be to create a dynamical field theory from our multiplets - as dynamical as it can be in two dimensions! For this we require a method of constructing Lagrangians and actions which are invariant under Lorentz transformations, translations and supersymmetry transformations. For reasons beyond the level of this introductory course, we shall also require scale invariance: our actions should be dimensionless (once $c = \hbar = 1$) without use of parameters, such as masses, which carry a dimension.

Translation, scale and Lorentz invariant actions are usually constructed by taking the integral over all space of a Lorentz invariant Lagrangian density with a scaling behaviour which compensates for that of the integral measure. In two dimensions, this could e.g. be something like

$$\mathcal{I} = \tfrac{1}{2} \int d^2x \, \partial_\mu A \, \partial^\mu A$$

with $A(x)$ a dimensionless field. The translational invariance comes about because the Lagrangian $\mathcal{L} = \frac{1}{2}\partial_\mu A \partial^\mu A$ transforms (like any field) as a divergence $\delta\mathcal{L} = \delta x^\mu \partial_\mu \mathcal{L} = \partial_\mu(\delta x^\mu \mathcal{L})$ under a constant translation $x^\mu \to x^\mu + \delta x^\mu$. The scale invariance comes about because d^2x has (mass) dimension +2 which is compensated by the two derivatives.

A Lagrangian which is a superfield will transform under an infinitesimal supersymmetry transformation with a constant spinor parameter ς_- as follows:

$$\delta\mathcal{L} = -i\varsigma_- r(Q_+)\, \mathcal{L} = -i\varsigma_- \frac{\partial}{\partial\theta_-} \mathcal{L} + \theta_- \partial_{++} \mathcal{L} \, .$$

The last term is just a translation which won't contribute under an $\int d^2x$. The first term, however, is actually a "translation in θ-space", $\theta_- \to \theta_- + \varsigma_-$. In order to construct an invariant, we must have a translation-invariant integration in θ-space.

The general "superfunction" $f = f_0 + \theta f_1$ transforms as $f \to (f_0 + \varsigma f_1) + \theta f_1 \equiv f_0' + \theta f_1$ under a tranlation $\theta \to \theta + \varsigma$. Hence, if the integral $\int d\theta \, f$ should not "see" ς, it must be independent of f_0. Furthermore, after the θ-variable has been integrated over, the result should not depend on it. Finally, the integral should be a linear functional. All of this is satisfied by the *Berezin integration*

$$\int d\theta\ (f_0 + \theta f_1) \equiv f_1 \ . \tag{4.1}$$

Notice that in our case of Berezin's integral, the integral measure $d\theta_-$ is not a Lorentz invariant. If f_1 is a scalar, then f_0 and with it the whole f must have a "-" Lorentz index.

We can summarize all of this into the following prescription for an action:

$$\mathcal{S} = \int d^2x\ d\theta_-\ \mathcal{L}_-(x,\theta)\ , \tag{4.2}$$

where $\mathcal{L}_-$ is a superfield with mass dimension 3/2. Note that, in contrast to the case of coordinates which are real numbers, the covariance properties of $d\theta$ are not those of the coordinates but those of the gradient (compare also Eq. (6.15) below). We could write

$$d\theta_- \equiv d(\theta_-) = (d\theta)_+ \tag{4.3}$$

to indicate this.

The simplest Lagrangian that we can construct like this for a scalar dimensionless superfield $\Phi(x)$ is

$$\mathcal{L}_-(x,\theta) = -\tfrac{1}{2}\ D_+\Phi\ \partial_{--}\Phi\ . \tag{4.4a}$$

In components, the Lagrangian $\mathcal{L}(x)$ which gives us the same action through $\mathcal{S} = \int d^2x\ \mathcal{L}(x)$ is

$$\mathcal{L}(x) = \tfrac{1}{2}\ \partial_{++}A\ \partial_{--}A + \tfrac{1}{2}\ \Psi_+\partial_{--}\Psi_+\ , \tag{4.4b}$$

and if we abandon light-cone notation, this becomes the familiar

$$\mathcal{L}(x) = \tfrac{1}{2}\ \partial_\mu A\ \partial^\mu A + \tfrac{1}{2}\ \bar{\Psi}\ \gamma^\mu\ \partial_\mu\ \Psi \tag{4.4c}$$

with Ψ the chiral spinor

$$\Psi = \begin{bmatrix} \Psi_+ \\ 0 \end{bmatrix} .$$

Another action can be constructed from a spinor superfield $\Lambda_-(x,\theta)$ of dimension $\tfrac{1}{2}$ by means of the superfield Lagrangian

$$\mathcal{L}_-(x,\theta) = -\tfrac{1}{2}\ \Lambda_- D_+ \Lambda_-\ , \tag{4.5a}$$

or in components (note that $F_{-+} \equiv F$ is a Lorentz scalar),

$$\mathcal{L}(x) = \tfrac{1}{2}\ \lambda_-\partial_{++}\lambda_- + \tfrac{1}{2}\ F^2\ , \tag{4.5b}$$

and in standard notation

$$\mathcal{L}(x) = \tfrac{1}{2}\,\bar{\lambda}\gamma^{\mu}\partial_{\mu}\lambda + \tfrac{1}{2}F^2\,, \tag{4.5c}$$

with λ the anti-chiral spinor

$$\lambda = \begin{bmatrix} 0 \\ \lambda_- \end{bmatrix}.$$

If we now count the degrees of freedom, we have the following ("off-shell") fields:

field:	number of real components:
$A(x)$	1 bosonic
$\psi_+(x)$	1 fermionic
$\lambda_-(x)$	1 fermionic
$F(x)$	1 bosonic

The same number of fermionic and bosonic components is an indicator for the supersymmetry present in the system.

The equations of motion which follow from the actions (4.4) and (4.5), respectively, are

$$\Box A = \gamma^{\mu}\partial_{\mu}\psi = 0 \quad \text{and} \quad \gamma^{\mu}\partial_{\mu}\lambda = F = 0\,, \tag{4.6a}$$

or in light-cone notation

$$\partial_{++}\partial_{--}A = \partial_{--}\psi_+ = 0 \quad \text{and} \quad \partial_{++}\lambda_- = F = 0\,. \tag{4.6b}$$

These have the solutions

$$A = A(x^+) + A(x^-)\;; \quad \psi_+ = \psi_+(x^+)\;; \quad \lambda_- = \lambda_-(x^-)\;; \quad F = 0\,, \tag{4.7}$$

and if we now count modes, we get

field:	right-moving modes:	left-moving modes:
$A(x)$	1 bosonic	1 bosonic
$\psi_+(x)$	1 fermionic	0 fermionic
$\lambda_-(x)$	0 fermionic	1 fermionic
$F(x)$	0 bosonic	0 bosonic

We see that in the right-moving sector the numbers of boson and fermion modes match, but in the left-moving sector they don't. This is typical for the (1,0) case where there is a Q_+ but no Q_- supersymmetry.

5. (1,1)-SUPERSYMMETRY

We saw that neither of our free models had a supersymmetry in the left-moving sector. If, however, we consider both together, we notice that the numbers of fermions and bosons match in the left-moving sector as well. Thus the question arises whether the sum of the two actions (4.4) and (4.5),

$$\begin{aligned} \mathcal{I} &= \tfrac{1}{2} \int d^2x \, (\partial_{++}A \, \partial_{--}A + i\psi_+\partial_{--}\psi_+ + i\lambda_-\partial_{++}\lambda_- + F^2) \\ &= \tfrac{1}{2} \int d^2x \, (\partial_\mu A \, \partial^\mu A + i\bar{\psi} \, \gamma^\mu\partial_\mu\psi + i\bar{\lambda} \, \gamma^\mu\partial_\mu\lambda + F^2) \, , \end{aligned} \tag{5.1}$$

does perhaps have full (1,1) supersymmetry. This is indeed the case: a second supersymmetry transformation is given by

$$\begin{aligned} [A,Q_-] &= i\lambda_- \; ; \quad \{\psi_+,Q_-\} = -F \\ \{\lambda_-,Q_-\} &= \partial_{--}A \; ; \quad [F,Q_-] = -i\partial_{--}\psi_+ \, . \end{aligned} \tag{5.2}$$

These transformations fulfill the algebra $Q_-Q_- = P_{--}$ and $\{Q_+,Q_-\} = 0$, as required. Moreover, the action (5.1) is invariant under Q_- transformations.

We can, of course, develop the full (1,1) tensor calculus from scratch: the algebra is

$$\{Q,\bar{Q}\} = 2\gamma^\mu P_\mu \quad \text{or} \quad \{Q,Q\} = -2\gamma^\mu C P_\mu \, , \tag{5.3}$$

with Q a Majorana spinor (Eq. (2.1) holds) but not chiral (Eq. (2.6) does not hold). The superspace is parametrized by the two real components of a Majorana spinor θ, and a scalar superfield is

$$\Phi(x,\theta) = A(x) + \bar{\theta}\psi(x) - \tfrac{1}{2}\bar{\theta}\theta F(x) \, , \tag{5.4}$$

with ψ now a two-component Majorana spinor. The supersymmetry transformations are given by

$$[\Phi,Q] = r(Q)\Phi \tag{5.5}$$

with

$$r(Q) \equiv i \frac{\partial}{\partial\bar{\theta}} - \gamma^\mu\theta\partial_\mu \, . \tag{5.6}$$

Together with $r(P_\mu) \equiv i\partial_\mu$, this represents the algebra (5.3). In components, Eq. (5.5) reads

$$[A,Q] = i\psi \; ; \quad \{\psi,Q\} = iCF - \gamma^\mu C \, \partial_\mu A \; ; \quad [F,Q] = -\gamma^\mu\partial_\mu\psi \, . \tag{5.7}$$

All of this is completely consistent with the (1,0) notation if we assume that the two-component spinors are

$$\varphi = \begin{bmatrix}\varphi_+ \\ \lambda_-\end{bmatrix} \; ; \quad Q = \begin{bmatrix}Q_+ \\ Q_-\end{bmatrix} \; ; \quad \theta = \begin{bmatrix}\theta_+ \\ \theta_-\end{bmatrix} . \tag{5.8}$$

There is, of course, also a spinorial superfield derivative, namely

$$D \equiv \frac{\partial}{\partial\bar{\theta}} - i\gamma^{\mu}\theta\partial_{\mu} . \tag{5.9}$$

The Berezin integral over all superspace is now

$$\int d^2\theta \equiv \int d\theta_+ \, d\theta_- . \tag{5.10}$$

With our conventions we get $\bar{\theta}\theta = 2i\theta_-\theta_+$ and thus

$$\int d^2\theta \; \bar{\theta}\theta = 2i . \tag{5.11}$$

Working out the details, we find that the action (5.1) is, in (1,1) notation:

$$\mathcal{I} = \frac{1}{4i} \int d^2x \, d^2\theta \; \bar{D\Phi} \, D\Phi . \tag{5.12}$$

The general form of a scale, Lorentz and (1,1) supersymmetry invariant action is

$$\mathcal{I} = \frac{1}{2i} \int d^2x \, d^2\theta \; \mathcal{L}(x,\theta) \tag{5.13}$$

where $\mathcal{L}$ is a Lorentz scalar superfield with (mass) dimension 1 so that $\mathcal{I}$ is dimensionless.

The only generalisation of the Lagrangian of Eq. (5.12) which is possible for a single scalar superfield is

$$\mathcal{L} = \tfrac{1}{2} \, g(\Phi) \; \bar{D\Phi} \; D\Phi \tag{5.14}$$

with $g(\Phi)$ an arbitrary dimensionless scalar function of the superfield Φ.

A further candidate for a Lagrangian that one could think of would be

$$\mathcal{L} = \tfrac{1}{2}\bar{D\Phi}\gamma_3 D\Phi$$

which is a pseudoscalar with all required properties. It is zero, however: we can write

$$\bar{D\Phi} D\Phi = C_{\alpha\beta} \, \bar{D}^{\alpha}\Phi \, \bar{D}^{\beta}\Phi \tag{5.15a}$$

$$\bar{D\Phi} \; \gamma_3 \; D\Phi = (\gamma_3 C)_{\alpha\beta} \bar{D}^{\alpha}\Phi \; \bar{D}^{\beta}\Phi \tag{5.15b}$$

and since $C = -C^T$ but $(\gamma_3 C) = (\gamma_3 C)^T$, the first expression can be non-zero but the second is always zero for classical fields, for which $\{\bar{D}^{\alpha}\Phi, \bar{D}^{\beta}\Phi\} = 0$.

We could consider *two or more superfields* ϕ^i and take a term

$$\mathcal{L}_{WZ} = \tfrac{1}{2}\, b_{ij}\, \overline{D\phi}^i\, \gamma_3\, D\phi^j \qquad (5.16)$$

into the Lagrangian, with b_{ij} some antisymmetric matrix. In components, the corresponding action reads

$$\mathcal{I}_{WZ} = -\tfrac{1}{2} \int d^2x\, b_{ij}\, \epsilon_{\mu\nu}\, \partial^\mu (A^i \partial^\nu A^j - \tfrac{i}{2}\, \bar{\psi}^i \gamma^\nu \psi^j)\ , \qquad (15.17)$$

which is, of course, the integral over a divergence and hence again zero. However, we are now in a position to write down the *most general* (1,1) *Lagrangian* for a set of scalar, dimensionless superfields ϕ^i:

$$\mathcal{L} = \tfrac{1}{2}\, g_{ij}(\phi)\, \overline{D\phi}^i\, D\phi^j + \tfrac{1}{2}\, b_{ij}(\phi)\, \overline{D\phi}^i\, \gamma_3\, D\phi^j\ . \qquad (5.18)$$

Here g_{ij} is a symmetric and b_{ij} an antisymmetric matrix of dimensionless Lorentz scalar functions of the superfields.

This is the Lagrangian for the (1,1) *supersymmetric σ-model*. The second term is called the *Wess-Zumino* term and is now non-zero since the integrand of (5.17) will not be a divergence if b_{ij} is not a constant.

6. REPARAMETRISATION INVARIANCE

The supersymmetric σ-model given by the Lagrangian (5.18) may under certain conditions have a very much larger symmetry than one might have expected.

Consider a general transformation of the superfields ϕ^i into each other

$$\phi^i \to \phi'^i(\phi)\ . \qquad (6.1)$$

Under this "general coordinate transformation" of the fields, we find that

$$D\phi^i \to D\phi^j\, \frac{\partial \phi'^i}{\partial \phi^j} \qquad (6.2)$$

by the chain rule. This is the transformation behaviour of a contravariant vector. If, therefore, g_{ij} and b_{ij} transform as covariant tensors, e.g.,

$$g_{ij}(\phi) \to g_{kl}(\phi)\, \frac{\partial \phi^k}{\partial \phi'^i}\, \frac{\partial \phi^l}{\partial \phi'^j}\ , \qquad (6.3)$$

then the Lagrangian (5.18) is actually invariant under the very large group of transformations (6.1). This section is dedicated to showing how this invariance manifests itself in the component version of the σ-model action with no Wess-Zumino term:

$$\mathcal{I} = \frac{1}{4i} \int d^2x\, d^2\theta\, g_{ij}(\phi)\, \bar{D}\phi^i\, D\phi^j \, . \tag{6.4}$$

First, we observe that whereas $\partial_\mu \phi^i$, $D\phi^i$ and $\bar{D}\phi^i$ are all contravariant vectors, second derivatives are not:

$$\partial_\mu D\phi^i \rightarrow \partial_\mu (D\phi^j \frac{\partial \phi'^i}{\partial \phi^j}) = (\partial_\mu D\phi^j) \frac{\partial \phi'^i}{\partial \phi^j} + D\phi^j (\partial_\mu \frac{\partial \phi'^i}{\partial \phi^j}) \, .$$

We must covariantise the derivative to make the second term go away. For this we first introduce the "Christoffel" connection, well known from General Relativity,

$$\Gamma_{ij}{}^k \equiv \tfrac{1}{2}\, g^{kl}(\partial_i g_{jl} + \partial_j g_{il} - \partial_l g_{ij}) \tag{6.5}$$

with g^{ij} the inverse of g_{ij} and

$$\partial_i g_{jl} \equiv \frac{\partial}{\partial \phi^i} g_{jl}(\phi) \, . \tag{6.6}$$

The transformation law of $\Gamma_{ij}{}^k$ is as in G.R.,

$$\Gamma_{ij}{}^k \rightarrow \frac{\partial \phi^m}{\partial \phi'^i} \frac{\partial \phi^n}{\partial \phi'^j} \left[\Gamma_{mn}{}^l \frac{\partial \phi'^k}{\partial \phi^l} - \partial_m \frac{\partial \phi'^k}{\partial \phi^n} \right] , \tag{6.7}$$

so that the covariant derivatives of tensors, typically

$$\mathcal{D}_i V_j \equiv \partial_i V_j - \Gamma_{ij}{}^k V_k \quad \text{and} \quad \mathcal{D}_i V^j \equiv \partial_i V^j + V^k \Gamma_{ik}{}^j \, , \tag{6.8}$$

are again tensors.

The definition (6.5) is equivalent to demanding that the Christoffel connection be symmetric (no torsion) and that the metric be covariantly constant:

$$\Gamma_{ij}{}^k = \Gamma_{ji}{}^k \tag{6.9}$$

$$\mathcal{D}_k g_{ij}(\phi) = 0 \, . \tag{6.10}$$

Defining connections ω_μ and ω_α through

$$\omega_{\mu i}{}^j \equiv \partial_\mu \phi^k\, \Gamma_{ki}{}^j \; ; \quad \omega_{\alpha i}{}^j \equiv \mathcal{D}_\alpha \phi^k\, \Gamma_{ki}{}^j \tag{6.11}$$

we find that they transform as

$$\omega_{\mu i}{}^j \rightarrow \frac{\partial \phi^m}{\partial \phi'^i} (\omega_{\mu m}{}^n \frac{\partial \phi'^j}{\partial \phi^n} - \partial_\mu \frac{\partial \phi'^j}{\partial \phi^m}) \tag{6.12}$$

$$\omega_{\alpha i}{}^j \rightarrow \frac{\partial \phi^m}{\partial \phi'^i} (\omega_{\alpha m}{}^n \frac{\partial \phi'^j}{\partial \phi^n} - \mathcal{D}_\alpha \frac{\partial \phi'^j}{\partial \phi^m})$$

and that the covariant derivatives

$$\mathcal{D}_\mu V_i \equiv \partial_\mu V_i - \omega_{\mu i}{}^j V_j \; ; \quad \mathcal{D}_\mu V^i \equiv \partial_\mu V^i + \omega_{\mu j}{}^i V^j \tag{6.13}$$

$$\mathcal{D}_\alpha V_i \equiv D_\alpha V_i - \omega_{\alpha i}{}^j V_j \; ; \quad \mathcal{D}_\alpha V^i \equiv D_\alpha V^i + \omega_{\alpha j}{}^i V^j$$

are vectors if V_i and V^i are. A consequence of Eq. (6.10) and of the definitions (6.11) is that also

$$\mathcal{D}_\mu g_{ij}(\phi) = \mathcal{D}_\alpha g_{ij}(\phi) = 0 \; . \tag{6.14}$$

How can we use all this to evaluate the action (6.4)? We first note, by looking at Eq.(3.12) and the definition (4.1) of the Berezin integral, that integrating over or differentiating with respect to an anticommuting variable is the same thing,

$$\int d\theta \; f(\theta) = \frac{\partial}{\partial\theta} f(\theta) \; . \tag{6.15}$$

In other words, we can replace the integration in the action by a differentiation. Under an integral over x-space, there is no difference between $\partial/\partial\bar{\theta}^\alpha$ and D_α since the $-i\gamma^\mu\theta\partial_\mu$ term in D_α does not contribute, and working out numerical factors, we find that the action (6.4) can also be written as

$$\mathcal{I} = -\frac{1}{8}\int d^2x \; \bar{D}D(g_{ij}(\phi)\;\bar{D}\phi^i D\phi^j) \; . \tag{6.16}$$

Since covariant derivatives obey the Leibniz rule just as ordinary ones do, we can evaluate this as

$$\mathcal{I} = \frac{1}{4}\int d^2x \; g_{ij}(\phi)\;(\;\bar{\mathcal{D}}^\alpha \bar{D}^\beta \phi^i \; \mathcal{D}_\alpha D_\beta \phi^j - \bar{D}^\alpha\phi^i \; \bar{\mathcal{D}}\,\mathcal{D}\,D_\alpha\phi^j \;) \; , \tag{6.17}$$

and the task remains to evaluate the multiple covariant derivatives. We find by explicit calculation

$$\mathcal{D}_\alpha D_\beta \phi^i = -i(\gamma^\mu C)_{\alpha\beta}\partial_\mu\phi^i + C_{\alpha\beta}\mathcal{F}^i \tag{6.18}$$

with

$$\mathcal{F}^i \equiv \tfrac{1}{2}\,\bar{\mathcal{D}}D\phi^i \tag{6.19a}$$

The superfield $\mathcal{F}^i$ is a contravariant vector under reparametrisations and can also be written as

$$\mathcal{F}^i = \tfrac{1}{2}\,\bar{D}D\phi^i + \tfrac{1}{2}\,\bar{D}\phi^k D\phi^l \Gamma_{kl}{}^i \; . \tag{6.19b}$$

We can use Eq. (6.18) as an intermediate result to evaluate $\bar{\mathcal{D}}\,\mathcal{D}\,D_\alpha\phi^i$ and find

$$\bar{\mathcal{D}}\,\mathcal{D}\,D_\alpha\phi^i = -i(\gamma^\mu\mathcal{D})_\alpha\partial_\mu\phi^i - \mathcal{D}_\alpha\mathcal{F}^i \; . \tag{6.20}$$

We observe first that

$$\mathcal{D}_\alpha\partial_\mu\phi^i = D_\alpha\partial_\mu\phi^i + \omega_{\alpha k}{}^i\partial_\mu\phi^k = D_\alpha\partial_\mu\phi^i + D_\alpha\phi^l\partial_\mu\phi^k\Gamma_{lk}{}^i = \mathcal{D}_\mu D_\alpha\phi^i \; , \tag{6.21}$$

using the symmetry of $\Gamma_{kl}{}^i$. We now have to evaluate $\mathcal{D}_\alpha \mathcal{F}^i$. We get

$$\tfrac{1}{2}\,\mathcal{D}_\alpha \bar{D}D\phi^i = \tfrac{1}{2}\,D_\alpha \bar{D}D\phi^i + \tfrac{1}{2}\,D_\alpha\phi^k \bar{D}D\phi^l \Gamma_{kl}{}^i$$

$$\tfrac{1}{2}\,\mathcal{D}_\alpha(\bar{D}\phi^k D\phi^l \Gamma_{kl}{}^i) = -\,\bar{D}^\beta\phi^k D_\alpha D_\beta\phi^l \Gamma_{kl}{}^i$$

$$+\tfrac{1}{2}\,\bar{D}\phi^k D\phi^l D_\alpha \Gamma_{kl}{}^i + \tfrac{1}{2}\,D_\alpha\phi^j \bar{D}\phi^k D\phi^l \Gamma_{kl}{}^m \Gamma_{jm}{}^i \,.$$

Using $D_\alpha D_\beta \phi^i = -i(\gamma^\mu C)_{\alpha\beta}\partial_\mu \phi^i + \tfrac{1}{2}\,C_{\alpha\beta}\,\bar{D}D\phi^i$, we get

$$-\,\bar{D}^\beta\phi^k D_\alpha D_\beta \phi^l \Gamma_{kl}{}^i = i(\gamma^\mu D)_\alpha \phi^k \omega_{\mu k}{}^i - \tfrac{1}{2}\,D_\alpha\phi^k \bar{D}D\phi^l \Gamma_{kl}{}^i \,.$$

A product of three D's cannot be fully antisymmetrized (there are only two spinor index values). Hence

$$D_\alpha D_\beta D_\gamma = D_\alpha D_\beta D_\gamma - \frac{1}{6}\,(D_\alpha D_\beta D_\gamma - D_\alpha D_\gamma D_\beta + \text{cyclic})$$

$$= \frac{1}{2}\,D_\alpha\{D_\beta, D_\gamma\} + \frac{1}{3}\,\{D_\alpha, D_\beta\}D_\gamma + \frac{1}{6}\,D_\gamma\{D_\alpha, D_\beta\} - \frac{1}{6}\,D_\beta\{D_\alpha, D_\gamma\} - \frac{1}{3}\,\{D_\alpha, D_\gamma\}D_\beta$$

and

$$D_\alpha \bar{D}D = 2i(\gamma^\mu D)_\alpha \partial_\mu \,. \tag{6.22}$$

Collecting our terms, we find the intermediate result

$$\mathcal{D}_\alpha \mathcal{F}^i = i(\gamma^\mu \mathcal{D}_\mu D)_\alpha \phi^i + \tfrac{1}{2}\,\bar{D}\phi^k D\phi^l\, D_\alpha\phi^j\, (\partial_j \Gamma_{kl}{}^i + \Gamma_{kl}{}^m \Gamma_{jm}{}^i) \,.$$

The term in brackets at the end must be a covariant expression; we expect that it has something to do with the Riemann tensor

$$R_{jkl}{}^i \equiv \partial_j \Gamma_{kl}{}^i - \partial_k \Gamma_{jl}{}^i - \Gamma_{jl}{}^m \Gamma_{km}{}^i + \Gamma_{kl}{}^m \Gamma_{jm}{}^i \,. \tag{6.23}$$

Indeed, we can use the completeness property of the γ-matrices, do Fierz rearrangements and get identities

$$2\gamma_3\psi^l(\bar{\psi}^k \gamma_3 \psi^j)\Gamma_{kl}{}^i = \gamma^\mu\psi^l(\bar{\psi}^k \gamma_\mu \psi^j)\Gamma_{kl}{}^i = -\,\psi^j(\bar{\psi}^k\psi^l)\Gamma_{kl}{}^i$$

which hold for any anticommuting Majorana spinors ψ^i because $\Gamma_{kl}{}^i$ is symmetric in k and l. We therefore have

$$\psi^j(\bar{\psi}^k\psi^l)\Gamma_{kl}{}^i = -\,\frac{2}{3}\,(\,\gamma^\mu\psi^l(\bar{\psi}^k\gamma_\mu\psi^j) + \gamma_3\psi^l(\bar{\psi}^k\gamma_3\psi^j)\,)\,\Gamma_{kl}{}^i$$

and, since both terms in brackets are antisymmetric in k and j, we get

$$\psi^j(\bar{\psi}^k\psi^l)\,(\partial_j\Gamma_{kl}{}^i + \Gamma_{kl}{}^m\Gamma_{jm}{}^i) = \frac{2}{3}\,(-\,\tfrac{1}{2}\gamma^\mu\psi^l(\bar{\psi}^k\gamma_\mu\psi^j) - \tfrac{1}{2}\gamma_3\psi^l(\bar{\psi}^k\gamma_3\psi^j))\,R_{jkl}{}^i.$$

We can now add a further term $-\tfrac{1}{2}\psi^l(\bar{\psi}^k\psi^j)$ to the terms in brackets (which won't change anything since it is symmetric in k and j) and find that the result is just the Fierz rearrangement of

$$\frac{2}{3}\,\psi^j(\bar{\psi}^k\psi^l)\,R_{jkl}{}^i \,.$$

We thus have the results

$$\mathscr{D}_\alpha \mathscr{F}^i = i(\gamma^\mu \mathscr{D}_\mu D)_\alpha \phi^i + \frac{1}{3}\bar{D}\phi^k D\phi^l D_\alpha \phi^j R_{jkl}{}^i(\phi) \tag{6.24}$$

$$\bar{\mathscr{D}}\,\mathscr{D}\,D_\alpha \phi^i = -\,2i(\gamma^\mu \mathscr{D}_\mu D)_\alpha \phi^i - \frac{1}{3} D_\alpha \phi^j \bar{D}\phi^k D\phi^l R_{jkl}{}^i(\phi)\,, \tag{6.25}$$

and the action (6.17) is now

$$\mathscr{I} = \tfrac{1}{4}\int d^2x\, g_{ij}(\phi)\,(2\partial_\mu \phi^i \partial^\mu \phi^j + 2\mathscr{F}^i \mathscr{F}^j \tag{6.26}$$

$$+2i\bar{D}\phi^i \gamma^\mu \mathscr{D}_\mu D\phi^j + \frac{1}{3}\bar{D}\phi^i D\phi^m \bar{D}\phi^k D\phi^l\, R_{mkl}{}^j)\,.$$

Since in the action only the θ-independent components contribute (the others are divergences), we can use

$$\phi^i = A^i + o(\theta)\ ; \qquad D_\alpha \phi^i = \psi^i + o(\theta) \tag{6.27}$$

$$\mathscr{F}^i = F^i + \tfrac{1}{2}\bar{\psi}^k \psi^l \Gamma_{kl}{}^i(A) + o(\theta) \equiv \tilde{F}^i + o(\theta)\,,$$

and write

$$\mathscr{L} = \frac{1}{2} g_{ij}\partial_\mu A^i \partial^\mu A^j + \frac{1}{2}\tilde{F}_i \tilde{F}^i + \frac{i}{2}\bar{\psi}_i \gamma^\mu \mathscr{D}_\mu \psi^i - \frac{1}{12}\bar{\psi}^i \psi^k\, \bar{\psi}^j \psi^l R_{klij}\,, \tag{6.28}$$

with

$$\mathscr{D}_\mu \psi^i = \partial_\mu \psi^i + \partial_\mu A^k \psi^l \Gamma_{kl}{}^i\,. \tag{6.29}$$

Here g_{ij}, $\Gamma_{kl}{}^i$ and R_{klij} are taken as functions of the ordinary fields A^i only, rather than of the superfields ϕ^i.

This is the final component form of the (1,1) supersymmetric σ-model Lagrangian without the Wess-Zumino term.

SUPERGRAVITY AS A GAUGE THEORY

Allen C. Hirshfeld

Institut für Physik der Universität Dortmund
Postfach 500 500, 4600 Dortmund 50, Fed. Rep. Germany

ABSTRACT

It is shown that in a certain sense Einstein gravity may be considered as the limit of a Yang-Mills type gauge theory with the anti-de-Sitter group as gauge group. Simple supergravity in four dimensions may then be constructed in a straightforward way as a supersymmetric extension of ordinary gravity.

The two major achievements of 20th century physics are quantum mechanics and general relativity. The outstanding problem of integrating these quite different theories into a unified world view has occupied many minds. Relativistic quantum field theory is a successful merger of quantum theory and special relativity, and it provides an accurate description of electromagnetic interactions on the microscopic scale in the theory of Quantum Electrodynamics. The strengths of the electromagnetic and the gravitational interactions, however, are so disparate that a succesful theory for one does not lead in any direct way to the theory for the other. In terms of energy, the electromagnetic processes occur at the scale of a few GeV at most, whereas gravity is characterized by the Planck mass of 10^{19} GeV. Nevertheless, when the Yang-Mills theories, which are generalizations of the Abelian gauge theory of QED, proved capable of dealing with the weak interactions up to energy scales of 100 GeV there seemed to be ground for optimism. Finally, when in the grand unified theories a unification of the strong, weak and electromagnetic forces at a scale of 10^{15} GeV was postulated, the discrepancy with the energy scale of gravity no longer seemed so formidable, and the time seemed ripe to try anew to unify gravity with the other fundamental interactions.

Another development in elementary particle physics also played an important role in the birth of supergravity. Ever since the introduction of internal symmetries in high energy physics, the goal of uniting internal and space-time symmetries had seemed a natural direction of progress. Various no-go theorems, however, blocked this line of advance, until it was realized that by taking the step from Lie groups, which until then had been the only groups used to describe symmetries in physics, to the so-called super Lie-groups, which involve a graded mathematical structure, the restrictions of the no-go theorems could be circumvented. Indeed, supersymmetry is necessary in order to unite particles of different spin into a single symmetry multiplet, and this in turn is a prerequisite for unifying gravity, mediated by the spin 2 gravitons, with the Yang-Mills theories, mediated by spin 1 gauge bosons.

Supergravity, which may be viewed as the supersymmetric extension of a Yang-Mills type gauge theory is, seen in this way, a natural candidate for a unified field-theory. Early investigations were indeed encouraging, for it turned out that the divergences in perturbative calculations, which had plagued all earlier attempts at a quantum theory of gravity, were in supergravity much less malignment. It was hoped for a time that the theory would turn out to be finite at all orders of pertubation theory. Today, the general view is not so simplistic, and supergravity is believed to be the low energy manifestation of a superstring theory on which, in turn, the expectation of finiteness is now concentrated.

In this review we do not treat the extended supergravity theories in higher dimensions, which really attempt a realistic unification of gravity with the nuclear forces. We restrict ourselves to simple supergravity in four dimensions, which describes a spin 2 field (the graviton) in interaction with a spin 3/2 field (the gravitino). Even this is no small triumph, for it is the first construction of a consistent field theory involving fields of spin greater than one.

We wish to describe supergravity as a supersymmetric gauge theory, which is in some way a natural extension of general relativity. The first problem with this concept is that Einstein gravity in its usual formulation is not a Yang-Mills type gauge theory in a direct sense. There have been various attempts to treat gravity as a gauge theory; an excellent review is given by Trautman in the Einstein Centennial Volume[1]. The approach we adopt here is dictated by the following considerations. Global supersymmetry is built on the supersymmetric extension of the Poincaré group, and it is clear that a supergravity theory must contain the Poincaré group as a local symmetry. The Poincaré group is the semi-direct product of the Lorentz group with the space-time translations, whereas in Yang-Mills theory we usually deal with simple Lie groups. It is therefore helpful to embed the Poincaré group in the anti-de-Sitter (AdS) group, which is simple. The Poincaré group may be regained by group contraction of the AdS group. For this reason we shall here treat Einstein gravity as the limit of a gauge theory with the AdS group as gauge group; supergravity will turn out to be a straightforward supersymmetric extension. This approach to supergravity was first used by Macdowell and Mansouri[2]. It allows various generalizations, in particular it was extensively used in the construction of superconformal gravity theories[3].

This article is divided into three sections. In the first we present a general introduction to Lie supergroups and superalgebras and discuss their relevance to physics. In particular we treat the Lie superalgebra osp(1/4) in some detail, since it is the supersymmetric extension of the Lie algebra of the AdS group, and it reduces, upon contraction, to the super-Poincaré algebra. The second section deals with formulations of gravity suitable for the present purposes, in particular with Cartan's approach to gravity theory and its relation to the usual tensor calculus. In the third section, it is shown how gravity may be treated as the limit of an AdS gauge theory, and the construction of the Supergravity Lagrangian is demonstrated.

Lie Superalgebras and Quantum Field Theory

A symmetry generator in quantum field theory is bilinear in the creation and annihilation operators and has the general form

$$G = \sum_{i,j} \int d^3p\, d^3q\, a_i^+(p)\, K_{ij}(p,q)\, a_j(q)\ . \tag{1.1}$$

Here p and q are 3-momenta, i and j additional quantum numbers, $a_j(q)$ the Fock space annihilation operator for a particle with momentum q and quantum number j, $a_i^+(p)$ the corresponding creation operator. $K_{ij}(p,q)$ is an integral kernel specific to the particular symmetry operator considered. For the Hamiltonian, for example,

$$K_{ij}(p,q) = \omega_p\, \delta_{ij}\, \delta(\bar{p}-\bar{q})\ , \tag{1.2}$$

where $\omega_p = \sqrt{p^2 + m^2}$, with m the mass of the particle under consideration. In an abbreviated notation the generator may be written as:

$$G = a^+\, K\, a\ . \tag{1.3}$$

The condition for G to be a symmetry generator is $[G,S] = 0$, where S is the scattering operator. If G_1 and G_2 are symmetry generators, which commute with S, so is $[G_1,G_2]$ (by the Jacobi identity), so the symmetry generators form a closed Lie algebra.

Until recently the symmetries of interest in elementary particle physics have been those of bosonic particles among themselves, or fermions among themselves. In these cases the creation and annihilation operators in the above formulas are either bosonic operators, satisfying canonical commutation relations, or fermionic operators, satisfying canonical anticommutation relations. In the case of supersymmetry we are interested in symmetries between bosonic and fermionic particles. In this case the symmetry generators involve in general an even component, relating bosons to bosons and fermions to fermions, and an odd component, relating bosons to fermions. Explicitly:

$$G = B + Q\ , \tag{1.4}$$

with

$$B = b^+\, K_{bb}\, b + f^+\, K_{ff}\, f \tag{1.5}$$

and

$$Q = f^+\, K_{fb}\, b + b^+\, K_{bf}\, f\ . \tag{1.6}$$

The usual symmetry generators are of the bosonic B-type, the Q-type are called supersymmetry generators and they relate particles with spins differing by 1/2.

The canonical commutation relations for the bosonic creation and annihilation operators are:

$$[b,b^+] = 1 \; ; \quad [b^+,b^+] = [b,b] = 0 \; . \tag{1.7}$$

The canonical anticommutation relations for the fermionic operators are:

$$\{f,f^+\} = 1 \; ; \quad \{f^+,f^+\} = \{f,f\} = 0 \; . \tag{1.8}$$

We assume that the bosonic and fermionic operators commute:

$$[b,f] = [b^+,f^+] = [b,f^+] = [b^+,f] = 0. \tag{1.9}$$

The generalized symmetry generators given in Eqs. (4), like the usual symmetry generators, form a closed algebraic system. We work this out in detail first for the case of two bosonic generators.

Let $$B^1 = b^+ K^1_{bb}\, b + f^+ K^1_{ff}\, f \; ; \quad B^2 = b^+ K^2_{bb}\, b + f^+ K^2_{ff}\, f \tag{1.10}$$

Then we calculate

$$[B^1,B^2] = (K^1_{bb})_{ij}(K^2_{bb})_{mn}\, [b^\dagger_i b_j, b^\dagger_m b_n] + (K^1{}_{ff})_{ij}(K^2_{ff})_{mn}\, [f^\dagger_i f_j, f^\dagger_m f_n]. \tag{1.11}$$

In this equation the repeated indices indicate an integration with respect to momenta and a summation with respect to other quantum numbers. The commutators in Eq. (11) are worked out according to:

$$[b^\dagger_i b_j, b^\dagger_m b_n] = b^\dagger_i [b_j, b^\dagger_m] b_n + b_m [b^\dagger_i, b_n] b_j = \delta_{jm} b^\dagger_i b_n - \delta_{in} b^\dagger_m b_j \; ,$$

$$[f^\dagger_i f_j, f^\dagger_m f_n] = f^\dagger_i \{f_j, f^\dagger_m\} f_n - f^\dagger_m \{f^\dagger_i, f_n\} f_j = \delta_{jm} f^\dagger_i f_m - \delta_{in} f^\dagger_m f_j \; . \tag{1.12}$$

The commutators again yield bilinear combinations of the creation and annihilation operators. Inserting Eq. (12) in (11) we find

$$[B^1,B^2] = B^3 \; ,$$

with

$$B^3 = b^\dagger K^3_{bb} b + f^\dagger K^3_{ff} f$$

and

$$K^3_{bb} = [K^1_{bb}, K^2_{bb}] \; , \qquad K^3_{ff} = [K^1_{ff}, K^2_{ff}] \; . \tag{1.13}$$

A similar calculation leads to the result:

$$[Q^1, B^2] = Q^3 , \qquad (1.14)$$

with

$$Q^3 = f^{\dagger}K^3_{fb}b + b^{\dagger}K^3_{bf}f$$

and

$$K^3_{fb} = K^1_{fb}K^2_{bb} - K^2_{ff}K^1_{fb} , \quad K^3_{bf} = K^1_{bf}K^2_{ff} - K^2_{bb}K^1_{bf} . \qquad (1.15)$$

However, if we try to calculate the commutator $[Q^1,Q^2]$, we find that the calculation analogous to that in Eq. (12) does *not* lead to expressions bilinear in the creation and annihilation operators. In order to obtain a closed algebraic system we must consider the *anticommutator* $\{Q^1,Q^2\}$. For this object we find

$$\{Q^1, Q^2\} = B^3 , \qquad (1.16)$$

with

$$B^3 = b^{\dagger}K^3_{bb}b + f^{\dagger}K^3_{ff}f$$

and

$$K^3_{bb} = K^1_{bf}K^2_{fb} + K^2_{bf}K^1_{fb} , \quad K^3_{ff} = K^1_{fb}K^2_{bf} + K^2_{fb}K^1_{bf} . \qquad (1.17)$$

A closed algebraic system of this type is an example of a structure referred to in the mathematical literature as a *graded Lie Algebra*; in contrast to a usual Lie algebra it involves both commutators and anticommutators. The particular structure we are here concerned with is a *Lie superalgebra*. We have demonstrated above that the occurence of Lie superalgebras in the description of the symmetry generators describing a symmetry involving both bosons and fermions is, in quantum field theory, a direct consequence of the canonical commutation and anticommutation relations between bosonic and fermionic creation and annihilation operators.

Early research on supersymmetry in particle physics was motivated by the seminal paper of Haag, Łopusański and Sohnius[4], in which it was shown how Lie superalgebras allow one to circumvent the no-go theorem of Coleman and Mandula[5]. This theorem demonstrates rigorously that the framework of quantum field theory allows no non-trivial combination of spacetime symmetry with the usual internal symmetries. The no-go theorem presented for a time a serious block for the development of unified field theories. In particular it is not possible to unify the gravitational force, mediated by the spin 2 gravitons, with other fundamental forces of Yang-Mills type, mediated by the spin 1 gluons and vector bosons, without recourse to supersymmetry.

As discussed in the previous subsection, a Lie superalgebra consists of a set of even (bosonic) generators B_i and odd (fermionic) generators Q_α, closed under commutation and anticommutation according to the rules

$$[B_i,B_j] = c_{ij}^k\, B_k\ ; \qquad [Q_\alpha,B_i] = s_{\alpha i}^\beta\, Q_\beta\ ; \qquad \{Q_\alpha,Q_\beta\} = \gamma_{\alpha\beta}^i\, B_i\ . \qquad (1.18)$$

The quantities c_{ij}^k, $s_{\alpha i}^\beta$, $\gamma_{\alpha\beta}^i$ are the *structure constants* of the Lie superalgebra. The c_{ij}^α and $s_{\alpha i}^\beta$ are antisymmetric in the lower indices; the $\gamma_{\alpha\beta}^i$ are symmetric. In addition, the structure constants are related by the *super Jacobi identities*, which follow from the equations below:

$$\text{(i)}\quad [[B_i,B_j],B_k] + [[B_j,B_k],B_i] + [[B_k,B_i],B_j] = 0\ ,$$

$$\text{(ii)}\quad [[Q_\alpha,B_i],B_j] + [[B_i,B_j],Q_\alpha] + [[B_j,Q_\alpha],B_i] = 0\ ,$$

$$\text{(iii)}\quad [\{Q_\alpha,Q_\beta\},B_i] - \{[Q_\beta,B_i],Q_\alpha\} + \{[B_i,Q_\alpha],Q_\beta\} = 0\ ,$$

$$\text{(iv)}\quad [\{Q_\alpha,Q_\beta\},Q_\gamma] + [\{Q_\beta,Q_\gamma\},Q_\alpha] + [\{Q_\gamma,Q_\alpha\},Q_\beta] = 0\ . \qquad (1.19)$$

The rule here is to permute the indices cyclically and sum, changing the sign if two odd generators are commuted.

Similarly to ordinary Lie algebras, every Lie superalgebra allows a representation in terms of matrices whose entries are structure constants: the *adjoint representation*[6]. This is given by block matrices of the form

$$B_i = \begin{bmatrix} A_i & 0 \\ 0 & D_i \end{bmatrix}, \qquad Q_\alpha = \begin{bmatrix} 0 & X_\alpha \\ Y_\alpha & 0 \end{bmatrix}, \qquad (1.20)$$

where A_i, D_i, X_α, Y_α are the following matrices:

$$(A_i)^j{}_k = c_{ik}^j\ ; \qquad (D_i)^\beta{}_\alpha = s_{i\alpha}^\beta\ ,$$

$$(X_\alpha)^\beta{}_i = s_{\alpha i}^\beta\ ; \qquad (Y_\alpha)^i{}_\beta = \gamma_{\alpha\beta}^i\ . \qquad (1.21)$$

We now demonstrate that these matrices indeed give a representation of the Lie superalgebra. We calculate

$$[B_i,B_j] = \begin{bmatrix} [A_i,A_j] & 0 \\ 0 & [D_i,D_j] \end{bmatrix} \qquad (1.22)$$

and

$$[A_i,A_j]^m{}_n = c_{ik}^m c_{jn}^k - c_{jk}^m c_{in}^k = c_{ij}^k c_{kn}^m = c_{ij}^k (A_k)^m{}_n\ . \qquad (1.23)$$

Here we have used the ordinary Jacobi identity for the bosonic structure constants, which follows from Eq. (19i). Eq. (23) just states that the matrices A_i form the adjoint representation of the Lie subalgebra spanned by the bosonic generators alone.

Continuing the calculation, we find for the second entry in the matrix in Eq. (22):

$$[D_i, D_j]^{\alpha}{}_{\beta} = s^{\alpha}_{i\gamma} s^{\gamma}_{j\beta} - s^{\alpha}_{j\gamma} s^{\gamma}_{i\beta} = c^{k}_{ij} s^{\alpha}_{k\beta} = c^{k}_{ij} (D_k)^{\alpha}{}_{\beta} \quad , \tag{1.24}$$

where we have now used the Jacobi identity which follows from Eq. (19ii). This shows that the D_i form a representation (in general reducible) of the bosonic Lie subalgebra. We have thus verified that the first relation in Eq. (18), $[B_i, B_j] = c^{a}_{ij} B_k$, is satisfied by the matrices given in Eqs. (20),(21).

A similar calculation shows that these matrices also satisfy the second relation in Eq. (18), $[Q_{\alpha}, B_i] = S^{\beta}_{\alpha i} Q_{\beta}$. Here we must use the Jacobi identity which follows from Eq. (19iii):

$$\gamma^{j}_{\rho\beta}(D_i)^{\rho}{}_{\alpha} + \gamma^{j}_{\rho\alpha}(D_i)^{\rho}{}_{\beta} - \gamma^{k}_{\alpha\beta}(A_i)^{j}{}_{k} = 0 \; . \tag{1.25}$$

This equation tells us that the $\gamma^{i}_{\alpha\beta}$ are the components of a tensor which is invariant with respect to the bosonic Lie subalgebra, where the Greek subscripts of $\gamma^{i}_{\alpha\beta}$ transform according to the D_i-representation of the Lie subalgebra and the Latin superscript according to the adjoint representation.

Finally, the verification of the last equation in (18), $\{Q_{\alpha}, Q_{\beta}\} = \gamma^{i}_{\alpha\beta} B_i$, involves the Jacobi identity which follows from (19iv):

$$\gamma^{i}_{\alpha\beta}(D_i)_{\rho}{}^{\tau} + \gamma^{i}_{\beta\rho}(D_i)_{\alpha}{}^{\tau} + \gamma^{i}_{\rho\alpha}(D_i)_{\beta}{}^{\tau} = 0 \; . \tag{1.26}$$

Since $\gamma^{i}_{\alpha\beta}$ is an invariant tensor of the bosonic Lie subalgebra and D_i a representation of this subalgebra, Eq. (26) can be fullfilled only if the Lie subalgebra possesses an invariant tensor and a representation related in this way. In other words, given a definite Lie algebra, it may be extended to a Lie superalgebra only if it possesses an invariant tensor $\gamma^{i}_{\alpha\beta}$ and a representation D_i satisfying Eq. (26).

Our demonstration that the matrices (20) constitute a representation of the Lie superalgebra (18) is now complete. In the course of this demonstration we have learned that the structure constants of the Lie superalgebra, far from being arbitrary, are largely determined by the Lie subalgebra spanned by the bosonic generators alone. We shall see in the continuation that the structure constants of the particular Lie superalgebras with which we shall be concerned satisfy these general requirements. In fact, using these requirements, the extension of a given Lie algebra to the Lie superalgebra which contains the given algebra as its bosonic subalgebra is almost immediate. In this procedure we must add to the given bosonic generators B_i the fermionic supersymmetry generators Q_α. The 2nd formula of Eq. (18), $[Q_\alpha, B_i] = s_{\alpha i}^{\ \beta} Q_\beta$, together with Eq. (24), tells us that the Q_α transform as a representation (in general reducible) of the Lie subalgebra. As a matter of fact, due to physical requirements we are usually concerned with Q_α which transform as irreducible (spinor) representations of the Lie subalgebra.

Supermatrices

Considered as an abstract algebraic system, the Lie superalgebra of Eq. (18) allows different representations. We have already seen two such representations: in terms of operators in Fock space, and, in the adjoint representation, in terms of ordinary finite-dimensional numerical matrices. In both cases the binary operation indicated in Eq. (18), in general called the graded Lie bracket, is realized for two even generators by means of the operator or matrix commutator, for two odd generators by means of the anticommutator. For many purposes still another representation of the Lie superalgebras is useful, in terms of the supermatrices[7] to be discussed below. These are matrices which have as entries even and odd elements of a Grassman algebra. For the purpose of representing Lie superalgebras they have the advantage that the Lie bracket is represented in *all* cases by a matrix commutator. The change of sign for the case of two odd generators is effected by the anticommuting Grassman variables.

Our main interest in considering supermatrices here is the fact that we shall be concerned with *supermatrix groups*, whose generators form Lie superalgebras, just as the generators of the ordinary matrix groups form ordinary Lie algebras. While for ordinary abstract Lie groups and Lie algebras the relation between group and algebra can be formulated in a completely general way, independant of the matrix representations, this cannot be done in a direct manner for the Lie superalgebras and their corresponding supergroups. Indeed, the general concept of a Lie supergroup is not easy to formulate, but the construction of the supermatrix groups is quite straightforward, as we shall see below.

In the same way that ordinary matrices represent linear transformations in an ordinary vector space, *supermatrices* represent linear transformations in a *graded* vector space. That is, the vectors of this space have both bosonic components, represented by ordinary numbers or even elements of a Grassman algebra, and fermionic components, represented by the odd elements of the Grassman algebra. Explicitly, a vector of the graded space may be written as

$$v = \begin{bmatrix} b \\ f \end{bmatrix} , \tag{1.26}$$

where b is a column vector with numerical or even Grassman entries, f a column vector with odd Grassman entries. A linear transformation on this space takes the form

$$\begin{bmatrix} b \\ f \end{bmatrix} \to \mathcal{M} \begin{bmatrix} b \\ f \end{bmatrix} = \begin{bmatrix} A & X \\ Y & D \end{bmatrix} \begin{bmatrix} b \\ f \end{bmatrix} , \tag{1.27}$$

where $\mathcal{M}$ is a supermatrix which may be written as indicated in block form. $A = \mathcal{M}_{bb}$ is called the boson-boson part of $\mathcal{M}$, $D = \mathcal{M}_{ff}$ the fermion-fermion part and $X = \mathcal{M}_{bf}$, $Y = \mathcal{M}_{fb}$ the mixed parts. A and D are matrices which have as entries even Grassman elements, X and Y have as entries odd Grassman elements.

The general linear group of ordinary matrices GL(N) consists of all the invertible matrices (matrices with non-vanishing determinant). In the same way all the non-singular supermatrices form the general linear supermatrix group. It is necessary, however, to discuss the concept of non-singularity here with some care, since the supermatrix blocks X and Y are nilpotent and hence not invertible.

It proves convenient to write the supermatrix $\mathcal{M}$ as a product of two simpler supermatrices:

$$\mathcal{M} = \mathcal{P}\mathcal{T} = \begin{bmatrix} A & 0 \\ Y & I \end{bmatrix} \begin{bmatrix} I & A^{-1}X \\ 0 & D-YA^{-1}X \end{bmatrix} . \tag{1.28}$$

Here A,D,X,Y are the matrix blocks of Eq. (27) and I is the ordinary unit matrix. The inverse supermatrix may then be written as

$$\mathcal{M}^{-1} = \mathcal{T}^{-1}\mathcal{S}^{-1} = \begin{bmatrix} I & -A^{-1}X(D-YA^{-1}X)^{-1} \\ 0 & (D-YA^{-1}X) \end{bmatrix} \begin{bmatrix} A^{-1} & 0 \\ YA^{-1} & I \end{bmatrix} = \begin{bmatrix} \cdot & \cdot \\ \cdot & (D-YA^{-1}X) \end{bmatrix} . \quad (1.29)$$

We see that $\mathcal{M}^{-1}$ exists if A^{-1} and $(D-YA^{-1}X)^{-1}$ do.

The supermatrices with which we are concerned may be decomposed into an "ordinary" part and a nilpotent part:

$$\mathcal{M} = \mathcal{M}_o + \mathcal{M}_n = (I + \mathcal{M}_n\mathcal{M}_o^{-1})\mathcal{M}_o . \quad (1.30)$$

The inverse of such a supermatrix may then be computed according to

$$\mathcal{M}^{-1} = \mathcal{M}_o^{-1}(I + \mathcal{M}_n\mathcal{M}_o^{-1})^{-1} = \mathcal{M}_o^{-1}(I - \mathcal{M}_n\mathcal{M}_o^{-1} + \tfrac{1}{2}(\mathcal{M}_n\mathcal{M}_o^{-1})^2 + \ldots) . \quad (1.31)$$

Since $\mathcal{M}_n$ is nilpotent the power series in Eq. (31) is actually finite, and hence $\mathcal{M}^{-1}$ always exists if $\mathcal{M}_0{}^{-1}$ does. In particular, $D-YA^{-1}X$ is invertible if D_0 is, because the term $YA^{-1}X$ is nilpotent. The general supermatrix $\mathcal{M}$ of Eq. (27) is invertible if A_0 and D_0 are.

Most of the calculations involving supermatrices go through in complete analogy to ordinary matrix calculations, except for two novelties in the supermatrix case: the definition of the trace and the transposition operation.

For supermatrices the appropriate quantity , which is analogous to the trace for ordinary matrices, is the *supertrace*, defined by

$$\mathrm{str}\mathcal{M} = \mathrm{tr}A - \mathrm{tr}D \quad (1.32)$$

in terms of the block matrices of Eq. (27). It is immediately checked that with this definition the important property

$$\mathrm{str}\mathcal{M}_1\mathcal{M}_2 = \mathrm{str}\mathcal{M}_2\mathcal{M}_1 \quad (1.33)$$

is preserved, which would not be the case if the seemingly more plausible plus sign were used in Eq. (32).

To define the transposition operation for supermatrices, let us reconsider the linear transformations in a graded vector space, given in Eq. (27). A vector $v = (b,f)$ is transformed by the action of the supermatrix $\mathcal{M}$ into $v' = (b',f')$, where

$$b' = Ab + Xf ,$$
$$f' = Yb + Df .$$

Remembering that X,Y and f are made up of odd Grassman variables, we see that

$$b'^T = b^TA^T - f^TX^T ,$$
$$f'^T = b^TY^T + f^TD^T .$$

Hence the equation $v'^T = v^T \mathcal{M}^T$ holds if we define the transposed supermatrix according to

$$\mathcal{M}^T = \begin{bmatrix} A^T & Y^T \\ -X^T & D^T \end{bmatrix} . \qquad (1.34)$$

Again, this definition guarantees the property:

$$(\mathcal{M}_1\mathcal{M}_2)^T = \mathcal{M}_2{}^T\mathcal{M}_1{}^T . \qquad (1.35)$$

For ordinary matrices the determinant may be defined by

$$\det\mathcal{M} = \exp\,(\mathrm{tr}\,\ln\mathcal{M}) ,$$

where
$$\ln\mathcal{M} = \sum_{k=1}^{\infty} \frac{(-1)^{k-1}}{k} (\mathcal{M}-I)^k .$$

For supermatrices the corresponding concept is the *superdeterminant* defined by

$$\mathrm{sdet}\mathcal{M} = \exp\,(\mathrm{str}\,\ln\mathcal{M}) . \qquad (1.36)$$

Given two supermatrices, $\mathcal{M}_1 = \exp \mathcal{N}_1$, $\mathcal{M}_2 = \exp \mathcal{N}_2$, we have, by the Baker-Hausdorff formula:

$$\mathcal{M}_1\mathcal{M}_2 = \exp\,(\mathcal{N}_1+\mathcal{N}_2+\text{commutators})$$

and hence
$$\mathrm{sdet}\mathcal{M}_1\mathcal{M}_2 = \exp \mathrm{str}(\mathcal{N}_1+\mathcal{N}_2) =$$
$$\exp\,(\mathrm{str}\mathcal{N}_1)\,\exp\,(\mathrm{str}\mathcal{N}_2) = \mathrm{sdet}\mathcal{M}_1\,\mathrm{sdet}\mathcal{M}_2 . \qquad (1.37)$$

The property (33) of the supertrace garantees that the supertrace of the commutators vanishes.

Using the factorization of $\mathcal{M}$ given in Eq. (28) we may calculate $\mathrm{sdet}\mathcal{M}$ as

$$\mathrm{sdet}\mathcal{M} = \mathrm{sdet}\mathcal{S}\;\mathrm{sdet}\mathcal{T} .$$

$$\mathrm{str}\,\ln\mathcal{S} = \mathrm{str}\,\sum \frac{(-1)^{k-1}}{k} \begin{bmatrix} A-I & 0 \\ Y & 0 \end{bmatrix}^k = \mathrm{str}\,\sum \frac{(-1)^{k-1}}{k} \begin{bmatrix} (A-I)^k & \cdot \\ \cdot & 0 \end{bmatrix}$$

$$= \mathrm{tr}\,\sum \frac{(-1)^{k-1}}{k} (A-I)^k = \mathrm{tr}\,\ln A .$$

$$\mathrm{str}\,\ln\mathcal{T} = \sum \frac{(-1)^{k-1}}{k} \begin{bmatrix} 0 & A^{-1}X-I \\ 0 & D-YA^{-1}X-I \end{bmatrix}^k = \mathrm{str}\,\sum \frac{(-1)^{k-1}}{k} \begin{bmatrix} 0 & x \\ x & (D-YA^{-1}X-I)^k \end{bmatrix}$$

$$= -\,\mathrm{tr}\,\ln(D-YA^{-1}X) .$$

Hence

$$\mathrm{sdet}\mathcal{M} = \det A\,\det(D-YA^{-1}X)^{-1} = \det\mathcal{M}_{bb}\,\det(\mathcal{M}^{-1})_{ff} . \qquad (1.38)$$

We see that $\mathrm{sdet}\mathcal{M} \neq 0$ if and only if A and $D - YA^{-1}X$ are invertible. Comparing with Eq. (29) we see that, just as for ordinary matrices, the criterion for the supermatrix $\mathcal{M}$ to be invertible is $\mathrm{sdet}\mathcal{M} \neq 0$.

In exactly the same way that all the ordinary non-singular matrices form a group, with matrix multiplication as the group operation, so do all the non-singular supermatrices. If the block matrix A in Eq. (27) is of dimension N and the matrix D of dimension M this group is denoted as the *superlinear group* SL(M/N).

The Orthosymplectic Groups and their Lie Superalgebras

The semi-simple Lie groups, with their associated Lie algebras, can be completely classified according to a scheme devised by Cartan. It turns out that there are essentially four infinite series of such groups, the so-called classical groups, each of which is a subgroup of the general linear group: the orthogonal groups O(N), of even and odd dimensions, the symplectic groups Sp(N) and the unitary groups U(N). In addition there are five exceptional groups.

Much work has been done in order to achieve a classification of the Lie superalgebras[8]. It turns out that the two main series of these superalgebras correspond again to subgroups of the superlinear group discussed in the previous section. They are the orthosympletic groups OSp(N/M) and the superunitary groups U(N/M). We shall be concerned here with the former series. In this series we shall find the super-Poincaré algebra and the super de-Sitter algebra. The latter series give rise, in a similar way, to the superconformal algebras which form the basis of conformal supergravity theories.

The orthosympectic groups are formed by those supermatrices of SL(N/M) which leave the bilinear form $z^T Hw$ invariant, where z and w are elements of a graded vector space:

$$z = \begin{bmatrix} x \\ \theta \end{bmatrix} \; ; \quad w = \begin{bmatrix} y \\ \chi \end{bmatrix} ,$$

and H is a block diagonal matrix

$$H = \begin{bmatrix} \eta & 0 \\ 0 & C \end{bmatrix} . \tag{1.39}$$

Here η is a symmetric $N \times N$ matrix, and C an antisymmetric $M \times M$ matrix, where M is an even number. The bilinear form is given explicitly by:

$$z^T H w = \sum_{i,j=1}^{N} x^i \eta_{ij} y^j + \sum_{\alpha,\beta=1}^{M} \theta^\alpha C_{\alpha\beta} \chi^\beta . \tag{1.40}$$

This is of course reminiscent of the condition for ordinary orthogonal matrices, which leave the first term in this expression invariant, and sympletic matrices, which leave the second term invariant. As usual, x and y are taken to have even Grassmanic entries, θ and χ odd entries.

The generators G of the Lie superalgebra osp(N/M) are just the infinitesimal parts of those Osp(N/M) supermatrices close to the identity: $M = I + G$. The invariance condition

$$z'^T H w' = z^T(I+G^T)H(I+G)w = z^T H w$$

yields for the generators the condition

$$G^T H + HG = 0 . \tag{1.41}$$

In block form the supermatrix G is

$$G = \begin{bmatrix} A & X \\ Y & D \end{bmatrix} ,$$

and the condition (41) becomes

$$A^T \eta + \eta A = 0 , \quad D^T C + CD = 0 , \quad X^T \eta - CY = 0 . \tag{1.42}$$

The even-even parts of the generators of OSp(M/N) are hence just the generators of the orthogonal groups $A \in o(N)$, the odd-odd parts the generators of the symplectic groups, $D \in sp(M)$.

The Lie Superalgebra osp(1/4) and the Supersymmetry Algebra

The Anti-De-Sitter Group SO(3,2)

We shall be particularly concerned in the following with the orthosymplectic group OSp(1/4), i.e. the case N=1 and M=4. In this subsection we consider first the bosonic Lie subalgebra which is the Lie algebra of the symplectic group Sp(4).

In 4 dimensions the 4×4 matrices D may be expanded in terms of the well-known Dirac bilinears $1, \gamma^a, \sigma^{ab}, \gamma^a\gamma^5$ and γ^5. The matrix C in Eq. (42) may be taken to be the antisymmetric charge conjugation matrix, which satisfies

$$C\gamma^a C^{-1} = -(\gamma^a)^T . \tag{1.43}$$

Eq. (42) now states that the matrices $CD = -D^T C = D^T C^T = (CD)^T$ are symmetric, hence they can be expanded in terms of the ten symmetric bilinears $C\gamma^a$ and $C\sigma^{ab}$, and the matrices D can be expanded in terms of the basis

$$M_{ab} = \sigma_{ab} = \tfrac{1}{4} [\gamma^a, \gamma^b] , \quad \Pi_a = \frac{1}{2} \gamma_a . \tag{1.44}$$

With this explicit representation for the generators the Lie algebra sp(4) may be calculated; the result is

$$[M_{ab},M_{cd}] = -\eta_{ac}M_{bd} + \eta_{ad}M_{bc} + \eta_{bc}M_{ad} - \eta_{bd}M_{ac} ,$$

$$[M_{ab},\pi_c] = \eta_{bc}\pi_a - \eta_{ac}\pi_b , \quad [\pi_a,\pi_b] = M_{ab} . \tag{1.45}$$

The indices a,b,c run from 1 to 4. We now introduce the indices m,n which run from 0 to 4, and relabel the generators as

$$M_{mn} = M_{ab} \quad \text{for } m=a, n=b ;$$

$$M_{mo} = -M_{om} = \pi_a \quad \text{for } m=a ; \quad M_{oo} = 0 . \tag{1.46}$$

The Lie algebra of Eq. (45) may then be compactly written as

$$[M_{mn},M_{rs}] = -\eta_{mr}M_{ns} + \eta_{ms}M_{nr} + \eta_{nr}M_{ms} - \eta_{ns}M_{mr} . \tag{1.47}$$

Here $\eta_{mn} = \eta_{ab} = \text{diag}(-1,+1,+1,+1)$ (for m=a, n=b), $\eta_{oo} = -1$ and $\eta_{ao} = \eta_{oa} = 0$ is the metric of the anti-de-Sitter group SO(3,2). So we have established that the Lie groups Sp(4) and SO(3,2) are locally isomorphic.

The physical interest of the anti-de-Sitter group emerges when we insert a length scale L into $\pi_a = LP_a$, so that P_a has the dimensions of a momentum in natural units ($\hbar$=c=1). Then the Lie algebra may be written as

$$[M_{ab},M_{cd}] = -\eta_{ac}M_{bd} + \cdots ,$$

$$[M_{ab},P_c] = \eta_{bc}P_a - \eta_{ac}P_b ,$$

$$[P_a,P_b] = L^{-2} M_{ab} . \tag{1.48}$$

In the limit $L \to \infty$ this is just the familiar Lie algebra of the Poincaré group. This process of passing from the group SO(3,2) to the Poincaré group $SO(3,1)\circledS R^4$ is called Wigner-Inönö contraction. While the Poincaré group is generally considered to be of more direct physical interest, the anti-de-Sitter group is more managable mathematically, since it is a simple Lie group, whereas the Poincaré group is a semi-direct product. The de-Sitter group SO(4,1) has also found use in the physical literature, but for our present purposes it is the anti-de-Sitter group SO(3,2) which is relevant, since the de-Sitter group has been shown to lead to the wrong sign for the energy in supergravity.

Besides the generators of the bosonic Lie subalgebra osp(1/4) contains the fermionic generators, which are of the general form

$$G = \begin{bmatrix} 0 & X \\ Y & 0 \end{bmatrix} \tag{1.49}$$

subject to the constraint of Eq. (42),

$$X^T = CY \quad ,$$

or

$$Y = C^{-1}X^T \ . \tag{1.50}$$

(for $N=\eta=1$). X and Y are matrices containing odd Grassman entries; we may write them as

$$X = \bar{\epsilon}_\alpha X^\alpha \ , \quad Y = \bar{\epsilon}_\alpha Y^\alpha \ , \tag{1.51}$$

where X^α, Y^α are *numerical* matrices, and the $\bar{\epsilon}_\alpha$ are anticommuting Grassman variables, which will later be seen to transform as charge-conjugate Majorana spinors. The X^α may be written in terms of a convenient basis as:

$$(X^\alpha)_\beta = \delta^\alpha_\beta$$

and the Y^α, from Eq. (50), as

$$(Y^\alpha)_\beta = (C^{-1})^{\beta\gamma}(X^\alpha)_\gamma = (C^{-1})^{\beta\alpha} \ . \tag{1.52}$$

The fermionic generators, written as numerical matrices, are

$$S^\alpha = \begin{bmatrix} 0 & X^\alpha \\ Y^\alpha & 0 \end{bmatrix} \ , \tag{1.53}$$

with the matrix elements $(S^\alpha)^o{}_\beta = \delta^\alpha_\beta$ and $(S^\alpha)^\beta{}_o = (C^{-1})^{\beta\alpha}$. The bosonic generators are

$$B_{mn} = \begin{bmatrix} 0 & 0 \\ 0 & M_{mn} \end{bmatrix} \ , \tag{1.54}$$

where the zero in the upper left-hand corner is just the generator of O(1).

Now that the explicit form of all the generators is given, it is straightforward to work out the algebra. The non-vanishing components of $[S^\alpha, B_{mn}]$ are

$$[S^\alpha, B_{mn}]^o{}_\beta = (S^\alpha)^o{}_\gamma (B_{mn})^\gamma{}_\beta = \delta^\alpha_\gamma (M_{mn})^\gamma{}_\beta = (M_{mn})^\alpha{}_\gamma \delta^\gamma_\beta = (M_{mn})^\alpha{}_\gamma (S^\gamma)^o{}_\beta$$

and

$$[S^\alpha,B_{mn}]^\beta{}_o = -(B_{mn})^\beta{}_\gamma(S^\alpha)^\gamma{}_o = -(M_{mn})^\beta{}_\gamma(C^{-1})^{\gamma\alpha} = -(M_{mn}C^{-1})^{\beta\alpha}$$

$$= -(M_{mn}C^{-1})^{\alpha\beta} = -(M_{mn})^\alpha{}_\gamma(C^{-1})^{\gamma\beta} = (M_{mn})^\alpha{}_\gamma(C^{-1})^{\beta\gamma} = (M_{mn})^\alpha{}_\gamma(S^\gamma)^\beta{}_o \ .$$

Here we have used the symmetry of the matrices $M_{mn}C^{-1}$ and the antisymmetry of C^{-1}. We have thus found the commutator

$$[S^\alpha,B_{mn}] = (M_{mn})^\alpha{}_\beta\, S^\beta \ . \tag{1.55}$$

For the fermionic generators, since we are now using a numerical matrix representation, the *anticommutator* must be calculated:

$$\{S^\alpha,S^\beta\}^\tau{}_\sigma = (S^\alpha)^\tau{}_o(S^\beta)^o{}_\sigma + (\alpha\leftrightarrow\beta) = (C^{-1})^{\lambda\alpha}\delta^\tau_\lambda\delta^\beta_\sigma + (\alpha\leftrightarrow\beta)$$

$$= \tfrac{1}{2}\,(C^{-1})^{\lambda\alpha}[(\gamma_a)^\tau{}_\sigma(\gamma^a)^\beta{}_\lambda - 2(\sigma_{ab})^\tau{}_\sigma(\sigma^{ab})^\beta{}_\lambda]$$

$$= \tfrac{1}{2}\,(\gamma^a C^{-1})^{\alpha\beta}(\gamma_a)^\tau{}_\sigma - (\sigma^{ab}C^{-1})^{\alpha\beta}(\sigma_{ab})^\tau{}_\sigma$$

$$= -\,(M^{mn}C^{-1})^{\alpha\beta}(M_{mn})^\tau{}_\sigma \ . \tag{1.56}$$

We have again used the symmetry of the matrices $M_{nm}C^{-1}$, and the minus sign in the last expression results from raising the indices on M_{mn}, since $M_{ao} = \pi_a$ but $M^{ao} = -\pi^a$. We have also used the Fierz rearrangement formula for the γ matrices:

$$\delta^\tau_\lambda\delta^\beta_\sigma = \tfrac{1}{4}\,[\delta^\tau_\sigma\delta^\beta_\lambda + (\gamma_a)^\tau{}_\sigma(\gamma^a)^\beta{}_\lambda - 2(\sigma_{ab})^\tau{}_\sigma(\sigma^{ab})^\beta{}_\lambda$$
$$- (\gamma_5\gamma_a)^\tau{}_\sigma(\gamma_5\gamma^a)^\beta{}_\lambda + (\gamma_5)^\tau{}_\sigma(\gamma_5)^\beta{}_\lambda] \ . \tag{1.57}$$

Terms like, e.g., $\gamma_5 C^{-1}$ cancel in the expression (56) because of the symmetrization in the indices $(\alpha\leftrightarrow\beta)$.

Collecting our results for the osp(1/4) superalgebra we see that the generators form a closed algebraic system, according to the graded scheme of Eq. (18):

$$[B_{mn},B_{rs}] = -\eta_{mr}B_{ns} + \eta_{ms}B_{nr} + \eta_{nr}B_{ms} - \eta_{ns}B_{mr} \ ,$$

$$[S^\alpha,B_{mn}] = (M_{mn})^\alpha{}_\beta S^\beta \ , \quad \{S^\alpha,S^\beta\} = -\,(M^{mn}C^{-1})^{\alpha\beta}\,B_{mn} \ . \tag{1.58}$$

The commutator of the bosonic generators is given by the structure constants of the Lie subalgebra SO(3,2) and the fermionic generators transform like a representation of this subalgebra.

If we write out the Lie superalgebra of Osp(1/4), as given in Eq. (58), in terms of the generators M_{ab} and Π_a, we find

$$[M_{ab},M_{cd}] = -\eta_{ac}M_{bd} + \cdots ; \quad [M_{ab},\Pi_c] = \eta_{bc}\Pi_a - \eta_{ac}\Pi_b; \quad [\Pi_a,\Pi_b] = M_{ab}$$

$$[S^\alpha,\Pi_a] = \tfrac{1}{2}(\gamma_a)^\alpha{}_\beta\, S^\beta ; \quad [S^\alpha,M_{ab}] = (\sigma_{ab})^\alpha{}_\beta\, S^\beta$$

$$\{S^\alpha,S^\beta\} = (\gamma^a C^{-1})^{\alpha\beta}\,\Pi_a - (\sigma^{ab}C^{-1})^{\alpha\beta}\,M_{ab} . \tag{1.59}$$

In this case the Wigner-Inönö contraction involves the rescaling $\Pi_a \rightarrow LP_a$ and $S^\alpha \rightarrow \sqrt{L}\,Q^\alpha$. Upon taking the limit $L \rightarrow \infty$ we find

$$[M_{ab},M_{cd}] = -\eta_{ac}M_{bd} + \cdots ; \quad [M_{ab},P_c] = \eta_{bc}P_a - \eta_{ac}P_b ; \quad [P_a,P_b] = 0 ,$$

$$[Q^\alpha,P_a] = 0 ; \quad [Q^\alpha,M_{ab}] = (\sigma_{ab})^\alpha{}_\beta\, Q^\beta ; \quad \{Q^\alpha,Q^\beta\} = (\gamma^a C^{-1})^{\alpha\beta}\, P_a . \tag{1.60}$$

This is the well-known *super-Poincaré algebra*, which forms the basis of supersymmetric physics. Its Lie subalgebra is that of the ordinary Poincaré group, the symmetry group of relativistic field theory. The *supersymmetry generators* Q^α are translation-invariant, but with respect to Lorentz transformations they transform like *Majorana spinors*. The last equation, involving the anticommutator of two supersymmetry generators, is often read as saying that the supersymmetry generators are the "*square root*" of the energy-momentum 4-vector. When used as a local gauge symmetry, this 4-vector is the generator of infinitesimal coordinate transformations, which is the reason that *the local version of supersymmetry is necessarily a theory of supergravity*. The super-Poincaré algebra (for massless particles the super-conformal algebra) is essentially the *unique* supersymmetric extension of the Poincaré algebra which is of physical relevance. This is discussed in detail in the paper of Haag, Łopuszański and Sohnius[4].

Einstein's Approach to Gravity

Tensor Calculus

In Einstein's approach to general relativity a central concept is the covariance of the laws of physics with respect to general coordinate transformations. An infinitesimal coordinate transformation may be written as:

$$x^{\mu'} = x^{\mu} + \epsilon^{\mu}(x) \ . \tag{2.1}$$

In this form we see that the transformation may also be considered to be a *local translation*.

Physical fields are described by functions on spacetime which have well-defined transformation properties with respect to general coordinate transformations. A *scalar field* transforms according to:

$$\Phi'(x') = \Phi(x) \ . \tag{2.2}$$

The infinitesimal form of this rule of transformation is:

$$\delta_{\epsilon}\Phi(x) = \Phi'(x) - \Phi(x) = \Phi(x-\epsilon) - \Phi(x) = -\epsilon^{\mu}\partial_{\mu}\Phi \ . \tag{2.2a}$$

A *contravariant vector field* transforms according to:

$$V^{\mu'}(x') = \left[\frac{\partial x^{\mu'}}{\partial x^{\nu}}\right] V^{\nu}(x) \ , \tag{2.3}$$

or

$$\delta_{\epsilon}V^{\mu} = -\epsilon^{\nu}\partial_{\nu}V^{\mu} + (\partial_{\nu}\epsilon^{\mu})V^{\nu} \ . \tag{2.3a}$$

A *covariant vector field* transforms as:

$$V_{\mu'}(x') = \left[\frac{\partial x^{\nu}}{\partial x^{\mu'}}\right] V_{\nu}(x) \ , \tag{2.4}$$

$$\delta_{\epsilon}V_{\mu} = -\epsilon^{\nu}\partial_{\nu}V_{\mu} - (\partial_{\mu}\epsilon^{\nu})V_{\nu} \ . \tag{2.4a}$$

A *mixed tensor field* transforms as:

$$M^{\mu'}{}_{\nu'}(x') = \left[\frac{\partial x^{\mu'}}{\partial x^{\alpha}}\right]\left[\frac{\partial x^{\beta}}{\partial x^{\nu'}}\right] M^{\alpha}{}_{\beta}(x) \ , \tag{2.5}$$

$$\delta_{\epsilon}M^{\mu}{}_{\nu} = -\epsilon^{\rho}\partial_{\rho}M^{\mu}{}_{\nu} + (\partial_{\rho}\epsilon^{\mu})M^{\rho}{}_{\nu} - (\partial_{\nu}\epsilon^{\rho})M^{\mu}{}_{\rho} \ . \tag{2.5a}$$

The generalization to higher-rank tensors should now be obvious.

The partial derivatives of a scalar field form the components of a covariant vector field (the gradient), but for vector fields and higher-rank tensors the partial derivatives do not transform like tensors. In order to form covariant derivatives of such fields it is necessary to introduce the *world connections* $\Gamma_{\mu\nu}^{\rho}(x)$. These quantities transform with respect to coordinate transformations according to the law:

$$\Gamma^{\rho'}_{\mu'\nu'}(x') = \left[\frac{\partial x^{\rho'}}{\partial x^{\alpha}}\right]\left[\frac{\partial x^{\beta}}{\partial x^{\mu'}}\right]\left[\frac{\partial x^{\gamma}}{\partial x^{\nu'}}\right]\Gamma^{\alpha}_{\beta\gamma}(x) - \left[\frac{\partial^2 x^{\rho'}}{\partial x^{\alpha}\partial x^{\beta}}\right]\left[\frac{\partial x^{\alpha}}{\partial x^{\mu'}}\right]\left[\frac{\partial x^{\beta}}{\partial x^{\nu'}}\right] . \quad (2.6)$$

The infinitesimal form of this law is:

$$\delta_{\epsilon}\Gamma^{\rho}_{\mu\nu} = -\partial_{\mu}\partial_{\nu}\epsilon^{\rho} - \epsilon^{\lambda}\partial_{\lambda}\Gamma^{\rho}_{\mu\nu} - (\partial_{\mu}\epsilon^{\lambda})\Gamma^{\rho}_{\lambda\nu} - (\partial_{\nu}\epsilon^{\lambda})\Gamma^{\rho}_{\mu\lambda} + (\partial_{\lambda}\epsilon^{\rho})\Gamma^{\lambda}_{\mu\nu} . \quad (2.6a)$$

The transformation law of the world connections is constructed in just such a way as to make the *world covariant derivative* of a contravariant vector field,

$$\mathcal{D}_{\mu}V^{\nu} = \partial_{\mu}V^{\nu} + \Gamma^{\nu}_{\mu\rho}V^{\rho} , \quad (2.7)$$

transform like a mixed tensor. The world covariant derivative of a covariant vector field is

$$\mathcal{D}_{\mu}V_{\nu} = \partial_{\mu}V_{\nu} - \Gamma^{\rho}_{\mu\nu}V_{\rho} , \quad (2.8)$$

and of a third-rank tensor, for example:

$$\mathcal{D}_{\lambda}M_{\mu\nu}{}^{\rho} = \partial_{\lambda}M_{\mu\nu}{}^{\rho} - \Gamma^{\tau}_{\lambda\mu}M_{\tau\nu}{}^{\rho} - \Gamma^{\tau}_{\lambda\nu}M_{\mu\tau}{}^{\rho} + \Gamma^{\rho}_{\lambda\tau}M_{\mu\nu}{}^{\tau} . \quad (2.9)$$

In all cases the world covariant derivatives transform like tensors.

For higher-order derivatives of the fields some care is required. For example $\mathcal{D}_{\mu}\mathcal{D}_{\nu}V^{\rho}$ is *not* a tensor, but the antisymmetrized derivative is. Direct calculation yields

$$[\mathcal{D}_{\mu},\mathcal{D}_{\nu}]V^{\rho} = R_{\mu\nu}{}^{\rho}{}_{\tau}V^{\tau} - T^{\tau}_{\mu\nu}\mathcal{D}_{\tau}V^{\rho} . \quad (2.10)$$

This equation is called the *generalized Ricci identity*. The quantities $R_{\mu\nu}{}^{\rho}{}_{\tau}$ and $T_{\mu\nu}{}^{\tau}$ are called the *curvature* and *torsion* components, respectively, and are given by

$$R_{\mu\nu}{}^{\rho}{}_{\tau} = \partial_{\mu}\Gamma^{\rho}_{\nu\tau} - \partial_{\nu}\Gamma^{\rho}_{\mu\tau} + \Gamma^{\rho}_{\mu\lambda}\Gamma^{\lambda}_{\nu\tau} - \Gamma^{\rho}_{\nu\lambda}\Gamma^{\lambda}_{\mu\tau} \quad (2.11)$$

and

$$T^{\rho}_{\mu\nu} = \Gamma^{\rho}_{\mu\nu} - \Gamma^{\rho}_{\nu\mu} . \quad (2.12)$$

It may be verified that the curvature and the torsion coefficients transform like the components of tensors, even though the world connections are not tensors.

In general relativity we work with *metric connections* which satisfy

$$\mathcal{D}_\mu g_{\nu\rho} = \partial_\mu g_{\nu\rho} - \Gamma^\tau_{\mu\nu} g_{\tau\rho} - \Gamma^\tau_{\mu\rho} g_{\tau\nu} = 0 \; . \tag{2.13}$$

Here $g_{\mu\nu}(x)$ is the symmetric non-degenerate *metric tensor* which measures distances between points of the (pseudo-) Riemannian manifold and defines the scalar product of two vectors at a space-time point. A vector V^α is said to be *parallel-transported* when $\mathcal{D}_\mu V^\alpha = 0$. The postulate (13) guaranties that the scalar product of two vectors is preserved under parallel transport:

$$\partial_\mu(g_{\alpha\beta}V^\alpha W^\beta) = \mathcal{D}_\mu(g_{\alpha\beta}V^\alpha W^\beta) = (\mathcal{D}_\mu g_{\alpha\beta})V^\alpha W^\beta + g_{\alpha\beta}(\mathcal{D}_\mu V^\alpha)W^\beta + g_{\alpha\beta}V^\alpha(\mathcal{D}_\mu W^\beta) = 0 \; .$$

By cyclically permuting the indices in Eq. (13) and adding the resulting equations, we may solve for the connections in terms of derivatives of the metric and the torsion:

$$\Gamma^\rho_{\mu\nu} = \Gamma^\rho_{\mu\nu}(g) - K^\rho_{\mu\nu} \; . \tag{2.14}$$

Here

$$\Gamma^\rho_{\mu\nu}(g) = \tfrac{1}{2}\, g^{\rho\tau}(\partial_\mu g_{\nu\tau} + \partial_\nu g_{\mu\tau} - \partial_\tau g_{\mu\nu}) \tag{2.15}$$

are the *Christoffel symbols* and

$$K_{\mu\nu\rho} = \tfrac{1}{2}\,(T_{\nu\rho\mu} - T_{\mu\nu\rho} - T_{\rho\mu\nu}) \tag{2.16}$$

is the *contortion tensor*. $g^{\mu\nu}$ is the inverse of $g_{\mu\nu}$ and is used, together with the metric tensor itself, to raise and lower world indices, e.g. $T_{\mu\nu\rho} = g_{\sigma\rho}T^\sigma_{\ \mu\nu}$. In the case of vanishing torsion the world connections equal the Christoffel symbols:

$$\Gamma^\rho_{\mu\nu} = \Gamma^\rho_{\mu\nu}(g) \quad . \tag{2.17}$$

The Einstein-Hilbert Action

The basic dynamical variables in Einstein's theory of gravitation are the ten independent components of the symmetric metric tensor. The torsion is taken to be zero *ab initio* so that the connections are given in terms of derivatives of the metric according to Eq. (17). The Einstein-Hilbert action is not of the Yang-Mills type; it is linear in the curvature:

$$S_E[g] = \frac{-1}{4\kappa^2}\int d^4x\, \sqrt{g}\, R(g) \; . \tag{2.18}$$

Here

$$g = |\det(g_{\mu\nu})| \quad \text{and} \quad \kappa^2 = 4\pi G \; ,$$

with G Newton's gravitational constant and R the *scalar curvature* - the contraction of the *Ricci tensor*:

$$R_{\mu\nu} = g^{\rho\sigma}R_{\mu\rho\nu\sigma} , \tag{2.19}$$

$$R = g^{\mu\nu}R_{\mu\nu} . \tag{2.20}$$

One may check that in the case of vanishing torsion the Ricci tensor is symmetric.

Varying the action with respect to the metric leads to the *Einstein field equations*

$$R^{\mu\nu}(g) - \tfrac{1}{2} g^{\mu\nu}R(g) = 0 . \tag{2.21}$$

The notation $R^{\mu\nu}(g)$ is meant to emphasize that the curvature tensor is here considered to be a function of the Christoffel-symbols (15), i.e. a function of the derivatives of the metric. Eqs. (21) are therefore a coupled set of 2nd-order partial differential equations, from which the metric is to be determined.

Cartan's Approach to Gravity

The Vierbein Formalism

In Einstein's approach to general relativity the basic dynamical variables are the components of the metric tensor $g_{\mu\nu}(x)$. In Cartan's approach the basic variables are the *vierbeins* and, in an intermediate stage, the *spin connections* to be defined below.

At every point x of a four-dimensional Riemannian manifold it is possible to find an orthonormal set of basis one-forms e^a (a=0,...,3) with

$$(e^a,e^b) = \eta^{ab} , \tag{2.22}$$

or, in terms of components,

$$g^{\mu\nu}e^a{}_\mu e^b{}_\nu = \eta^{ab} . \tag{2.22a}$$

Since the e^a are linearly independent the 4×4 matrix $e^a{}_\mu$ is non-singular, and we can introduce an inverse matrix $e_a{}^\mu$ such that

$$e^a{}_\mu e_a{}^\nu = \delta_\mu{}^\nu ; \quad e^a{}_\mu e_b{}^\mu = \delta^a_b . \tag{2.23}$$

With the inverse matrices Eq. (22) may be written as

$$g^{\mu\nu} = \eta^{ab} e_a{}^{\mu} e_b{}^{\nu} . \tag{2.24}$$

This equation may be interpreted to say that "the vierbeins are the square root of the metric". It is, however, important to note that this relation between the vierbeins and the metric is not unique. A local Lorentz transformation of the vierbeins is

$$e^a{}_{\mu}(x) = \Lambda^a{}_b(x)\, e^b{}_{\mu}(x) , \tag{2.25}$$

where

$$\eta_{ab}\, \Lambda^a{}_c(x)\, \Lambda^b{}_d(x) = \eta_{cd} \tag{2.26}$$

is the defining property of a Lorentz transformation. The inverse vierbeins transform as

$$e_a{}^{\mu}(x) = e_b{}^{\mu}(x)\, [\Lambda^{-1}(x)]^b{}_a . \tag{2.27}$$

Inserted in Eq. (24), this transformation leaves the metric $g^{\mu\nu}$ invariant. Hence, while in the metric description of gravitation the essential ingredient is covariance with respect to general coordinate transformations (i.e. local translations), in the vierbein formalism we demand covariance with respect to both *local translations* and *local Lorentz transformations*.

Eq. (24) may be rewritten as a relation between the vierbeins and the inverse vierbeins:

$$e_a{}^{\mu} = \eta_{ab}\, g^{\mu\nu}\, e^b{}_{\nu} . \tag{2.28}$$

In general, the Latin *Lorentz indices* of tensors are raised and lowered by use of the Minkowski metric η_{ab} and its inverse η^{ab}, the Greek *world indices* by use of the Riemannian metric $g_{\mu\nu}$ and its inverse $g^{\mu\nu}$.

An infinitesimal local Lorentz transformation is of the form

$$\Lambda(x) = 1 + \tfrac{1}{2}\, \lambda_{ab}(x)\, M^{ab} , \tag{2.29}$$

where the antisymmetric matrices M^{ab} are the generators of the Lorentz group, and the λ_{ab} are likewise antisymmetric. The factor $\frac{1}{2}$ in formulas like Eq. (29), which involve a summation over antisymmetric index pairs, is introduced in order to eliminate the redundancy of including identical terms in the sum, e.g. $\lambda_{12}M^{12} = \lambda_{21}M^{21}$. In the vector representation the generators are given by:

$$(M^{ab})^c{}_d = \eta^{ac}\, \delta^b_d - \eta^{bc}\, \delta^a_d . \tag{2.30}$$

Hence contravariant Lorentz 4-vectors transform according to

$$\delta_\lambda V^a = (1 + \tfrac{1}{2}\lambda_{cd} M^{cd})^a{}_b V^b - V^a = \lambda^a{}_b V^b \tag{2.31}$$

and covariant vectors according to

$$\delta_\lambda V_a = \lambda_a{}^b V_b , \tag{2.31a}$$

so that the scalar product $V^a W_a$ is invariant.

When describing vector and tensor quantities the tensor formalism and the vierbein formalism are mathematically equivalent (see the next subsection), but for the description of the interaction of gravitation with spinor fields the vierbein formalism is essential. With respect to coordinate transformations spinor fields are indistinguishable from world scalars: $\delta_\epsilon \psi = -\epsilon^\mu \partial_\mu \psi$. However, whereas scalars are invariant with respect to local Lorentz transformations, spinors transform according to the rule:

$$\delta_\lambda \psi = \tfrac{1}{2}\lambda_{ab}\,\sigma^{ab}\,\psi , \tag{2.32}$$

where

$$\sigma^{ab} = \tfrac{1}{4}[\gamma^a, \gamma^b] \tag{2.33}$$

are the Lorentz group generators in the spinor representation. The adjoint spinor $\bar\psi = \psi^+\gamma^0$ transforms according to

$$\delta_\lambda \bar\psi = -\tfrac{1}{2}\lambda_{ab}\bar\psi\,\sigma^{ab} , \tag{2.34}$$

so that the quantity $\bar\psi\psi$ is a Lorentz scalar.

In order to achieve a Lorentz covariant derivative, the *spin connection* $\omega_{\mu ab}$ is introduced. With respect to coordinate transformations it transforms like an ordinary covariant world vector. However, with respect to local Lorentz transformations it has the characteristic inhomogeneous transformation law of a gauge field:

$$\delta_\lambda \omega_{\mu ab} = -\partial_\mu \lambda_{ab} + \lambda_a{}^c \omega_{\mu cb} + \lambda_b{}^c \omega_{\mu ac} . \tag{2.35}$$

The last two terms in this equation are just those expected for the transformation of a tensor with two covariant Lorentz indices. The *Lorentz covariant derivative* may then be defined:

$$D_\mu = \partial_\mu + \tfrac{1}{2}\omega_{\mu ab}M^{ab} . \tag{2.36}$$

It is clear from this relation that the $\omega_{\mu ab}$ are, like the Lorentz group generators M^{ab}, antisymmetric in the indices a and b.

The Lorentz covariant derivative acts on spinors according to the spinor representation of the generators:

$$D_\mu \Psi = \partial_\mu \Psi + \tfrac{1}{2} \omega_{\mu ab} \sigma^{ab} \Psi . \qquad (2.37)$$

It acts on vectors according to the vector representation:

$$D_\mu V^a = \partial_\mu V^a + \omega_\mu{}^a{}_b V^b , \qquad (2.38)$$

$$D_\mu V_a = \partial_\mu V_a + \omega_{\mu a}{}^b V_b . \qquad (2.38a)$$

It is easily seen that, with the transformation law for the spin connection of Eq. (35), $D_\mu V^a$ is a world vector and a Lorentz vector, $D_\mu \Psi$ is a world vector and a Lorentz spinor.

In order to get a tensor structure for the higher derivatives it is again necessary to antisymmetrize. The *Ricci identity* here takes the form

$$[D_\mu, D_\nu] = \tfrac{1}{2} R_{\mu\nu ab} M^{ab} , \qquad (2.39)$$

with the curvature tensor

$$R_{\mu\nu ab} = \partial_\mu \omega_{\nu ab} - \partial_\nu \omega_{\mu ab} + \omega_{\mu a}{}^c \omega_{\nu cb} - \omega_{\nu a}{}^c \omega_{\mu cb} . \qquad (2.40)$$

It is easily verified that the curvature tensor transforms like a world- and Lorentz-tensor.

The *torsion* is conveniently defined in this formalism as

$$T_{\mu\nu}{}^a = D_\mu e_\nu{}^a - D_\nu e_\mu{}^a . \qquad (2.41)$$

It transforms like a world tensor and a Lorentz vector. The curvature tensor and the derivative of the torsion are related by the *1st Bianchi identity*:

$$R_{[\mu\nu\rho]}{}^a = - D_{[\rho} T_{\mu\nu]}{}^a , \qquad (2.42)$$

where $R_{\mu\nu\rho}{}^a = e^b{}_\rho R_{\mu\nu b}{}^a$. In this equation the brackets signify that the enclosed indices are to be cyclically permuted and the resulting terms summed. The *2nd Bianchi identity* involves the derivatives of the curvature tensor:

$$D_{[\mu} R_{\nu\rho]ab} = 0 . \qquad (2.43)$$

The torsion may be decomposed into a part involving the ordinary partial derivatives of the vierbeins and a part involving the spin connections:

$$T_{\mu\nu}{}^a = \Omega_{\mu\nu}{}^a + \omega_\mu{}^a{}_b e^b{}_\nu - \omega_\nu{}^a{}_b e^b{}_\mu , \qquad (2.44)$$

with

$$\Omega_{\mu\nu}{}^{a} = \partial_{\mu}e_{\nu}{}^{a} - \partial_{\nu}e_{\mu}{}^{a} . \tag{2.44a}$$

Multiply Eq. (44) by $e_{a\rho}$, cyclically permute the indices $[\mu\nu\rho]$ and add the resulting equations. This procedure allows us to solve for the spin connection, and we find

$$\omega_{\mu ab} = \omega_{\mu ab}(e) + K_{\mu ab} , \tag{2.45}$$

with

$$\omega_{\mu ab}(e) = \tfrac{1}{2}\, e_{a}{}^{\rho}\, e_{b}{}^{\nu} \left(\Omega_{\nu\rho}{}^{c}\, e_{c\mu} - \Omega_{\mu\nu}{}^{c}\, e_{c\rho} - \Omega_{\rho\mu}{}^{c}\, e_{c\nu} \right) \tag{2.45a}$$

and

$$K_{\mu ab} = \tfrac{1}{2}\, e_{a}{}^{\rho}\, e_{b}{}^{\nu} \left(e_{c\rho}\, T_{\mu\nu}{}^{c} - e_{c\mu}\, T_{\nu\rho}{}^{c} + e_{c\nu}\, T_{\rho\mu}{}^{c} \right) . \tag{2.45b}$$

In the absence of torsion obviously $\omega_{\mu ab} = \omega_{\mu ab}(e)$.

Relating the Vierbein- to the Tensor-Calculus

As long as we restrict ourselves to the consideration of vector and tensor quantities (i.e. excluding spinors) the vierbein and tensor calculi are mathematically equivalent. In this subsection we demonstrate how this correspondence works.

In the tensor calculus all quantities are expanded in terms of the "natural basis" associated with the coordinates x^{μ}. A one-form, for example, is written as $V = V_{\mu} dx^{\mu}$. Because of the transformation law

$$dx^{\mu'} = \left[\frac{\partial x^{\mu'}}{\partial x^{\nu}} \right] dx^{\nu} \tag{2.46}$$

the quantities V_{μ} form the components of a covariant world vector. In the vierbein formalism all quantities are expanded in terms of the orthonormal basis provided by the vierbeins: the one-form given above is now written as $V = V_{a}e^{a}$, and the quantities V_{a} form the components of a covariant Lorentz vector.

The vierbein coefficients $e^{a}{}_{\mu}$ relate the orthonormal basis to the natural basis:

$$e^{a} = e^{a}{}_{\mu}\, dx^{\mu} . \tag{2.47}$$

The inverse relation is

$$dx^{\mu} = e^{a}\, e_{a}^{\ \mu} . \tag{2.47a}$$

From these relationships we immediately see that the vierbein coefficients relate world vectors to Lorentz vectors and vice versa:

$$V_a = e_a^{\ \mu} V_{\mu} ; \qquad V^a = e^a_{\ \mu} V^{\mu} ,$$

$$V_{\mu} = e^a_{\ \mu} V_a ; \qquad V^{\mu} = e_a^{\ \mu} V^a . \tag{2.48}$$

In general the vierbein coefficients turn a Lorentz index of a tensor into a world index and vice versa. For example, given the transformation properties of the torsion in the vierbein formalism it is clear that the quantity

$$T_{\mu\nu}^{\ \ \rho} = e_a^{\ \rho}\, T_{\mu\nu}^{\ \ a} \tag{2.49}$$

transforms like a world tensor. We shall see below that it is identical to the torsion tensor used in the tensor calculus.

The spin connection is *not* a Lorentz tensor, it is related to the world connection by the equation appropriate for relating gauge potentials:

$$\Gamma_{\mu\nu}^{\ \ \rho} = e^{a\rho}\, \partial_{\mu} e_{a\nu} + e^{a\rho}\, \omega_{\mu a}^{\ \ \ b}\, e_{b\nu} , \tag{2.50}$$

or equivalently:

$$\Gamma_{\mu\nu}^{\ \ \rho} = e^{a\rho}\, D_{\mu} e_{a\nu} . \tag{2.51}$$

Using the fact that the spin connection transforms with respect to coordinate traansformations like an ordinary world vector we may verify, using Eq. (50), the inhomogeneous transformation law for the world connection given in Eq. (6).

Using Eq. (51) in the expression relating the torsion to the vierbein derivatives, Eq. (41), we may verify that this quantity is indeed related to the torsion tensor of the tensor calculus according to the relation given in Eq. (49). It then becomes obvious that the contortion, as introduced in the vierbein formalism in Eq. (45), is related to the contortion tensor of Eq. (16) according to

$$K_{\mu\nu\rho} = K_{\mu ab}\, e^a_{\ \nu}\, e^b_{\ \rho} . \tag{2.52}$$

Eq. (51) may also be used to relate world-covariant and Lorentz-covariant derivatives. For example:

$$\mathcal{D}_{\mu} V^{\nu} = e_a^{\ \nu}\, D_{\mu} V^a . \tag{2.53}$$

The same prescription also works for higher-rank tensors, for example the world covariant derivative given in Eq. (9) may be given in terms of Lorentz-covariant derivatives by

$$\mathcal{D}_\lambda M_{\mu\nu}{}^{\rho} = e^a{}_\mu\, e^b{}_\nu\, e_c{}^\rho\, D_\lambda M_{ab}{}^c \; . \tag{2.54}$$

This result may be used to transform the Ricci identity of the vierbein formalism, Eq. (39), into the generalized Ricci identity of the tensor calculus, Eq. (10). In the course of this calculation we verify that the curvature introduced in the vierbein formalism by Eq. (40) is related to the curvature tensor of Eq. (11) in the expected fashion:

$$R_{\mu\nu\rho\tau} = R_{\mu\nu ab}\, e^a{}_\rho\, e^b{}_\tau \; . \tag{2.55}$$

Note that the world connection obtained from the spin connection by Eq. (50) is automatically a metric connection, where the metric is related to the vierbeins by Eq. (24), since

$$\mathcal{D}_\mu g_{\nu\rho} = e^a{}_\nu\, e^b{}_\rho\, D_\mu (e_a{}^\sigma\, e_b{}^\tau\, g_{\sigma\tau}) = e^a{}_\nu\, e^b{}_\rho\, D_\mu \eta_{ab} = 0. \tag{2.56}$$

Einstein-Cartan Gravity

In Cartan's approach to gravity the basic dynamical variables are the vierbeins and the spin connections. The torsion is not set *ab initio* to zero. The Einstein action, written in terms of the above variables, is

$$S_E = -\frac{1}{4\pi\kappa^2}\int d^4x\; e\; R(\omega) \; , \tag{2.57}$$

where

$$e = \det(e^a{}_\mu) \tag{2.58}$$

and

$$R(\omega) = R_{\mu\nu ab}(\omega)\, e^{a\mu}\, e^{b\nu} \; . \tag{2.59}$$

Here $R_{\mu\nu ab}(\omega)$ is the curvature associated with the spin connection ω according to Eq. (40).

The action may also be written in an alternative form, which is sometimes useful:

$$S_E = \frac{1}{16\pi\kappa^2}\int d^4x\; \epsilon^{\mu\nu\lambda\rho}\, \epsilon_{abcd}\, e^c{}_\lambda\, e^d{}_\rho\, R_{\mu\nu}{}^{ab} \; . \tag{2.60}$$

Here ϵ_{abcd} is the totally antisymmetric invariant Lorentz tensor with $\epsilon^{0123} = 1$. It is related to the tensor $\epsilon^{\mu\nu\lambda\rho}$ according to

$$\epsilon^{\mu\nu\lambda\rho} = e\, e_a{}^{\mu}\, e_b{}^{\nu}\, e_c{}^{\lambda}\, e_d{}^{\rho}\, \epsilon^{abcd} \; . \tag{2.61}$$

This equation follows immediately from the definition of the determinant. The equivalence of the expressions (57) and (60) is an immediate consequence of the identity

$$\epsilon^{abcd}\, \epsilon_{cdef} = -2(\delta^a_e \delta^b_f - \delta^a_f \delta^b_e) \; . \tag{2.62}$$

The change in the action induced by a variation of the spin connection is most easily calculated by starting from the form given in Eq. (60). The Palatini identity here takes the form

$$\delta R_{\mu\nu ab} = D_{\mu}(\delta\omega_{\nu ab}) - D_{\nu}(\delta\omega_{\mu ab}) \tag{2.63}$$

and we find

$$\delta S_E = \frac{1}{8\kappa^2} \int d^4x\, \epsilon^{\mu\nu\lambda\rho}\, \epsilon_{abcd}\, e^c{}_{\lambda}\, e^d{}_{\rho}\, D_{\mu}(\delta\omega_{\nu}{}^{ab}) \; . \tag{2.64}$$

Since ϵ^{abcd} is an invariant Lorentz tensor and $\epsilon^{\mu\nu\lambda\rho}$ a scalar density the integration by parts may be simply performed; the result is

$$\delta S_E = -\frac{1}{4\kappa^2} \int d^4x\, \epsilon^{\mu\nu\lambda\rho}\, \epsilon_{abcd}\, e^c{}_{\lambda}\, (D_{\mu} e^d{}_{\rho})\, \delta\omega_{\nu}{}^{ab} \; . \tag{2.65}$$

The condition for this variation to vanish is

$$D_{\mu} e^a{}_{\nu} - D_{\nu} e^a{}_{\mu} = T_{\mu\nu}{}^a = 0 \; . \tag{2.66}$$

Hence the condition for vanishing torsion here results as an Euler-Lagrange equation of the Einstein action. The spin connection is no longer an independent dynamical variable, it is given in terms of the vierbeins by Eq. (45a), $\omega_{\mu ab} = \omega_{\mu ab}(e)$.

The variation of the action with respect to the vierbein is most easily calculated by starting from the form of the action given in Eq. (57). We need the variations

$$\delta e^{a\mu} = -\, e^{a\rho}\, \delta e_{c\rho}\, e^{c\mu} \tag{2.67}$$

and

$$\delta e = e\, e^{a\mu}\, \delta e_{a\mu} \; . \tag{2.68}$$

Eq. (67) is just a consequence of the definition of $e^{a\mu}$ as the inverse matrix to $e_{a\mu}$. Eq. (68) is most easily calculated from the expression of a determinant as

$$\ln \det M = \mathrm{tr} \ln M \ . \tag{2.69}$$

We can now calculate the variation of the action, the result is

$$\delta S_E = \frac{1}{2\kappa^2} \int d^4x \ e \ [\ R^{a\mu}(\omega) - \tfrac{1}{2} e^{a\mu} R(\omega) \ ] \ \delta e_{a\mu} \tag{2.70}$$

and the Euler-Lagrange equations are

$$R^{a\mu}(\omega) - \tfrac{1}{2} e^{a\mu} R(\omega) = 0 \ . \tag{2.71}$$

Here

$$R^{a\mu}(\omega) = e^{a\lambda} \ R_{\lambda\rho cd}(\omega) \ e^{c\mu} \ e^{d\rho}$$

and

$$R(\omega) = e_{a\mu} \ R^{a\mu}(\omega) \ .$$

This last is the same expression as that given in Eq. (59). Note that the condition $\omega = \omega(e)$ implies $\Gamma = \Gamma(g)$. Hence contracting Eq. (71) with the vierbein $e_a{}^{\nu}$ leads back to the Einstein field equations as given in Eq. (21).

For pure gravity we have hence demonstrated the equivalence of the Cartan and the Einstein formulations. For our present purposes the Cartan formulation is advantageous, since (i) it allows the inclusion of spinor fields coupled to gravity, which is important in the supersymmetric context, and (ii) it lends itself more easily to the formulation of gravity as a gauge theory, since the gauge potentials (the spin connections and the vierbeins) are treated as the basic dynamical variables. This is the subject of the following section.

Gauge Theories

The General Formalism

A gauge theory always involves some Lie group G which acts as a local symmetry group. Associated with this group is a Lie algebra (later superalgebra) whose generators satisfy the commutation relations

$$[G_B, G_C] = f^A_{BC}\, G_A \,. \tag{3.1}$$

Here $A,B,C = 1,\ldots,\dim G$ and f^A_{BC} are the *structure constants* of the group. When we consider Lie superalgebras the commutators must be replaced in the appropriate places by anticommutators, otherwise the following considerations go through unchanged.

The *adjoint representation* of the Lie algebra is realized by letting the generators act on elements of the Lie algebra $X = X^A G_A$ according to

$$\delta_B X = [G_B, X] = X^C\, [G_B, G_C] = f^A_{BC}\, X^C\, G_A \,. \tag{3.2}$$

We shall also write this relation as

$$\delta_B X^A = G_B(X^A) = f^A_{BC}\, X^C \,. \tag{3.2a}$$

From here we see that in the adjoint representation the generators may be represented by matrices whose entries are the structure constants:

$$[G_B]^A{}_C = f^A_{BC} \,. \tag{3.3}$$

We have already used this representation in our discussion of the adjoint representation of a Lie superalgebra in Section 1.

A gauge theory also involves, as the basic dynamical variable, a gauge potential

$$\mathcal{A}_\mu = \mathcal{A}^A_\mu\, G_A \,, \tag{3.4}$$

where μ is a spacetime index. With the gauge potential a covariant derivative may be formed

$$D_\mu = \partial_\mu + \mathcal{A}_\mu \,. \tag{3.5}$$

The gauge potential has a characteristic transformation law with respect to gauge transformations:

$$\delta_\epsilon A_\mu^A = - D_\mu \epsilon^A = - \partial_\mu \epsilon^A - A_\mu^B G_B(\epsilon^A) = - \partial_\mu \epsilon^A - f_{BC}^A A_\mu^B \epsilon^C . \quad (3.6)$$

Here $\epsilon = \epsilon^A G_A$ characterizes the (infinitesimal) gauge transformation. The second term in this transformation law would be characteristic of a quantity transforming according to the adjoint representation, the gauge transformation involves in addition the inhomogeneous derivative term. The gauge transformation law of the potential is such that if a physical field ϕ^A transforms, say, according to the adjoint representation

$$\delta_\epsilon \phi^A = f_{BC}^A \epsilon^B \phi^C$$

then so does its covariant derivative:

$$\delta_\epsilon (D_\mu \phi^A) = \partial_\mu (\delta_\epsilon \phi^A) + f_{BC}^A (\delta_\epsilon A_\mu^B) \phi^C + f_{BC}^A A_\mu^B (\delta_\epsilon \phi^C) = f^A{}_{BC} \epsilon^B (D_\mu \phi^C) . \quad (3.7)$$

Associated with the gauge potential is a *field strength* or *curvature tensor*:

$$[D_\mu, D_\nu] = R_{\mu\nu}{}^A G_A . \quad (3.8)$$

The components $R_{\mu\nu}{}^A$ may be calculated by inserting the expression (5) for the covariant derivative, the result is:

$$R_{\mu\nu}{}^A = \partial_\mu A_\nu^A - \partial_\nu A_\mu^A + f_{BC}^A A_\mu^B A_\nu^C . \quad (3.9)$$

The curvature satisfies the *Bianchi identity*:

$$D_{[\mu} R_{\nu\sigma]}{}^A = 0. \quad (3.10)$$

The variation of the curvature induced by an arbitrary variation of the gauge potential is given by the *Palatini identity*:

$$\delta R_{\mu\nu}{}^A = D_\mu \delta A_\nu^A - D_\nu \delta A_\mu^A . \quad (3.11)$$

In particular, for a gauge transformation:

$$\delta_\epsilon R_{\mu\nu}{}^A = - [D_\mu, D_\nu] \epsilon^A = - R_{\mu\nu}{}^B G_B \epsilon^A = f_{BC}^A \epsilon^B R_{\mu\nu}{}^C . \quad (3.12)$$

The *dynamics* of a physical system is given by a specification of the *action*, in a gauge theory the action is of the *Yang-Mills* type, i.e. quadratic in the field strength:

$$S = \int d^4x \, R_{\mu\nu}{}^A R_{\lambda\sigma}{}^B Q_{AB} Q^{\mu\nu\lambda\sigma} . \quad (3.13)$$

The coefficients Q_{AB} and $Q^{\mu\nu\lambda\sigma}$ may be taken to be independent of the spacetime variable x. They must possess, however, specific transformation properties, as we shall see immediately, in order to gaurantee the invariance of the theory with respect to its physical symmetries. The tensor $Q^{\mu\nu\lambda\sigma}$ is of course antisymmetric in the indices (μ,ν) and (λ,σ).

The variation of the action with respect to an arbitrary variation of the gauge potential may be calculated with the help of the Palatini identity, Eq. (11). The result is

$$\delta S = 2 \int d^4x \, (\delta R_{\mu\nu}{}^{A}) \, R_{\lambda\sigma}{}^{B} \, Q_{AB} \, Q^{\mu\nu\lambda\sigma} = 4 \int d^4x \, (D_{\mu}\delta A_{\nu}^{A}) \, R_{\lambda\sigma}{}^{B} \, Q_{AB} \, Q^{\mu\nu\lambda\sigma} \, . \tag{3.14}$$

In particular, for a gauge transformation:

$$\delta_{\epsilon} S = 2 \int d^4x \, f_{CD}^{A} \, \epsilon^{C} \, R_{\mu\nu}{}^{D} \, R_{\lambda\sigma}{}^{B} \, Q_{AB} \, Q^{\mu\nu\lambda\sigma} \, . \tag{3.15}$$

We shall repeatedly have occasion to recall these general results in the treatment of the various special cases discussed in the following sections.

Yang-Mills Theory

In Yang-Mills theory proper the symmetry group is compact and semi - simple. In this case the symmetric Killing form of the group is positive- (or, depending on the conventions chosen, negative-) definite; it is given by

$$g_{AB} = \mathrm{tr}\,(G_A G_B) = (G_A)^{C}{}_{D}(G_B)^{D}{}_{C} = f_{AD}^{C} \, f_{BC}^{D} \, . \tag{3.16}$$

The Killing form may be used, like a metric, to raise and lower group indices, and the quantities

$$f_{ABC} = g_{AD} \, f_{BC}^{D} \tag{3.17}$$

are totally antisymmetric. For the SU(N) groups the generators may be normalized so that

$$g_{AB} \sim \delta_{AB} \, . \tag{3.18}$$

In the action it is natural to choose $Q_{AB} = g_{AB}$. The Minkowski metric on spacetime is considered to be given and fixed, and it is again natural to choose

$$Q^{\mu\nu\lambda\sigma} = \tfrac{1}{2} \left(\eta^{\mu\lambda} \, \eta^{\nu\sigma} - \eta^{\nu\lambda} \, \eta^{\mu\sigma} \right) , \tag{3.19}$$

which respects the various symmetry properties of the coefficients. The gauge invariance of the action then follows from Eq. (15):

$$\delta_\epsilon S_{YM} = 2 \int d^4x \, f_{BCD} \, \epsilon^C \, R_{\mu\nu}{}^D \, R^{B\mu\nu} = 0 \, , \tag{3.20}$$

because of the antisymmetry of the coefficients f_{BCD} in the indices (B,D).

Topological Invariants

In theories of gravitation the spacetime metric is not considered to be given *a priori*. In treating gravitational theories as gauge theories we shall choose $Q^{\mu\nu\lambda\rho} = \epsilon^{\mu\nu\lambda\rho}$, where $\epsilon^{\mu\nu\lambda\rho}$ is the tensor density which is an invariant of the Lorentz group. The variation of a Yang-Mills type action, with this choice of coefficients, with respect to an arbitrary variation of the gauge potential yields a surprising result. The calculation is most easily done by using Eq. (14):

$$\begin{aligned} \delta S &= 4 \int d^4x \, (D_\mu \delta A_\mu^A) \, R_{\lambda\sigma}{}^B \, Q_{AB} \, \epsilon^{\mu\nu\lambda\sigma} \\ &= 4 \int d^4x \, \epsilon^{\mu\nu\lambda\sigma} \, \{ D_\mu(\delta A_\nu^A \, R_{\lambda\sigma}{}^B \, Q_{AB}) - \delta A_\nu^A (D_\mu R_{\lambda\sigma}{}^B) Q_{AB} - A_\nu^A \, R_{\lambda\sigma}{}^B (D_\mu Q_{AB}) \} \end{aligned}$$

The 2nd term in the last expression vanishes because of the Bianchi identity of Eq. (10). Ignoring a surface term in the 1st integral, the calculation may be continued by inserting the explicit expression for the covariant derivative, Eq. (5),

$$\begin{aligned} \delta S &= 4 \int d^4x \, \epsilon^{\mu\nu\lambda\sigma} \, A_\mu^D \, \{ G_D \, (\delta A_\nu^A) \, R_{\lambda\sigma}{}^B + \delta A_\nu^A \, G_D \, (R_{\lambda\sigma}{}^B) \} \, Q_{AB} \\ &= 4 \int d^4x \, \epsilon^{\mu\nu\lambda\sigma} \, A_\mu^D \, \{ f_{DC}^A \, \delta A_\nu^C \, R_{\lambda\sigma}{}^B + f_{DC}^B \, \delta A_\nu^A \, R_{\lambda\sigma}{}^C \} \, Q_{AB} \\ &= 4 \int d^4x \, \epsilon^{\mu\nu\lambda\sigma} \, A_\mu^D \, R_{\lambda\sigma}{}^B \, \delta A_\nu^C \, \{ f_{DC}^A \, Q_{AB} + f_{DB}^A \, Q_{CA} \} \\ &= 4 \int d^4x \, \epsilon^{\mu\nu\lambda\sigma} \, A_\mu^D \, R_{\lambda\sigma}{}^B \, \delta A_\nu^C \, G_D \, (Q_{CB}) \, . \end{aligned} \tag{3.21}$$

Here we have used the fact that the generators act on the indices of Q_{CB} according to the rule for the adjoint representation. These indices must transform in this way to ensure the invariance of the Yang-Mills type action, see Eqs. (12) and (13).

If the tensor Q_{AB} is now additionally taken to be an *invariant tensor* of the symmetry group then $G_C(Q_{AB}) = 0$ for all values of the index C, i.e. the group generators annihilate an invariant tensor because the group transformations themselves, which involve exponentials of the generators, leave such tensors invariant. In this case Eq. (21) tells us that the variation of the action with repect to an *arbitrary* variation of the gauge potential vanishes. This means that the action itself is a *topological invariant*. In fact, with the choices we have made here the action is proportional to the *Euler number*, and the calculation we have just gone through is a pedestrian derivation of the *Gauss-Bonnet* theorem of classical differential geometry. For further reading concerning the connections between gauge theories and differential geometry the review article of Eguchi, Gilkey and Hanson[9] may be recommended.

Gravity as a Gauge Theory

The Lorentz Group as a Local Symmetry

We have seen that the theory of gravity, in the Cartan formulation, involves as local symmetries the Lorentz group and the translations. We shall first consider the Lorentz group.

The six generators of the Lorentz group (3 rotations and 3 boosts) are labelled by an antisymmetric index pair (a,b), a,b = 0,1,2,3, which can take six distinct values. To adapt the general notation of the previous subsection to this case the group index used there will here be replaced by the index pair (a,b), so that the generators there denoted by G_A are here written as M_{AB}. These generators satisfy the following commutation relations:

$$[M_{ab},M_{cd}] = -\eta_{ac}M_{bd} + \eta_{ad}M_{bc} + \eta_{bc}M_{ad} - \eta_{bd}M_{ac} \; . \tag{3.22}$$

The structure constants are defined by

$$[M_{ab},M_{cd}] = \tfrac{1}{2}\, f^{ef}_{ab,cd}\, M_{ef} \tag{3.23}$$

and we read off:

$$f^{ab}_{cd,ef} = \{[\eta_{cf}\delta^a_d\delta^b_e - (e\leftrightarrow f)] - [c\leftrightarrow d]\} - \{(a\leftrightarrow b)\} \tag{3.24}$$

When the Lorentz group is used as the local symmetry group of a gauge theory it is associated with a gauge potential, denoted in the previous subsection by A^A_μ and here as $\omega_\mu{}^{ab}$. The covariant derivative is now

$$\overset{\circ}{D}_\mu = \partial_\mu + \tfrac{1}{2}\,\omega_\mu{}^{ab}\, M_{ab} \tag{3.25}$$

(the o over the D is used to distinguish this covariant derivative from that of the anti-de-Sitter group, to be introduced in the following subsection). The properties of this gauge potential and its associated curvature may all be straightforwardly derived by using the general formulas of the previous subsection and inserting the explicit expression for the structure constants, as we now demonstrate.

The transformation of the potentials with respect to a local Lorentz transformation is given by Eq. (6), which here becomes

$$\delta_\lambda \omega_\mu{}^{ab} = - \overset{\circ}{D}_\mu \lambda^{ab} = - \partial_\mu \lambda^{ab} - \tfrac{1}{4} f^{ab}{}_{cd,ef}\, \lambda^{cd}\, \omega_\mu{}^{ef}$$

$$= - \partial_\mu \lambda^{ab} - \lambda_a{}^c \omega_{\mu cb} - \lambda_b{}^c \omega_{\mu ac} \,. \tag{3.26}$$

This is seen to agree with the transformation property of the Cartan spin connection, as given in Eq. (2.35).

The curvature associated with the gauge potential is given by Eq. (8), which for this case is the same as the Ricci identity of Eq. (2.39). The components of the curvature tensor are given in Eq. (9),

$$\overset{\circ}{R}_{\mu\nu}{}^{ab} = \partial_\mu \omega_\nu{}^{ab} - \partial_\nu \omega_\mu{}^{ab} + \tfrac{1}{4} f^{ab}{}_{cd,ef}\, \omega_\mu{}^{cd} \omega_\nu{}^{ef}$$

$$= (\partial_\mu \omega_\nu{}^{ab} + \omega_\mu{}^{ac} \omega_{\nu c}{}^{b}) - (\mu \leftrightarrow \nu) \tag{3.27}$$

in agreement with Eq. (2.40). The Bianchi identity (10) is identical to Eq. (2.43) and the Palatini identity (11) is Eq. (2.63). Eq. (12) for the gauge transformation of the curvature is here

$$\delta_\lambda \overset{\circ}{R}_{\mu\nu}{}^{ab} = \tfrac{1}{4} f^{ab}{}_{cd,ef}\, \lambda^{cd}\, \overset{\circ}{R}_{\mu\nu}{}^{ef} = \lambda^a{}_c \overset{\circ}{R}_{\mu\nu}{}^{cb} + \lambda^b{}_c \overset{\circ}{R}_{\mu\nu}{}^{ac} \,, \tag{3.28}$$

which just says that the curvature transforms like a tensor with respect to Lorentz transformations.

We see that the Cartan spin connections may be identified with the gauge potentials of a gauge theory which involves the Lorentz group as a local symmetry, and then all the properties of the spin connection and its associated curvature follow automatically. However, the Einstein-Cartan theory of gravity includes, besides the spin connections, the vierbeins as dynamical variables. Hence a gauge theory based on the Lorentz group as a local symmetry, with no additional structure, cannot reproduce Einstein-Cartan gravity. Various possibilities have been investigated for introducing this additional structure. In the fiber bundle approach one makes use of the "soldering form", or canonical one-form, which is available on the frame bundle over spacetime reduced to the pseudo-orthogonal Lorentz frame bundle. For a review of this approach, and some others which have been studied in the literature, see Trautman[1]. For our purposes the most convenient method for introducing this additional structure is to enlarge the Lorentz group SO(3,1) to the anti-de-Sitter group SO(3,2). This is the subject of the following subsection.

The Anti-de-Sitter Group as a Local Symmetry

The anti-de-Sitter group SO(3,2) contains the Lorentz group SO(3,1) as a subgroup; the generators of the Lorentz group are a subset of the generators of the anti-de-Sitter group. For these generators the commutation relations and structure constants have been given in the previous subsection. The new generators of the anti-de-Sitter group are Π_a, $a = 1,..,4$, and their commutators with the Lorentz generators and among themselves are

$$[M_{ab}, \Pi_c] = f_{ab,c}^{\ \ \ \ \ d}\, \Pi_d$$

$$[\Pi_a, \Pi_b] = \tfrac{1}{2}\, f_{a,b}^{\ \ \ cd}\, M_{cd} \ . \qquad (3.29)$$

The structure constants in these equations may be read off from Eqs. (1.45), they are

$$f_{ab,c}^{\ \ \ \ \ d} = \eta_{bc}\delta^d_a - \eta_{ac}\delta^d_b \ ,$$

$$f_{a,b}^{\ \ \ cd} = \delta^c_a\delta^d_b - \delta^d_a\delta^c_b \qquad (3.30)$$

The components of the gauge potential associated with the new generators are designated by $e^a_{\ \mu}$, so the covariant derivative is

$$D_\mu = \partial_\mu + \tfrac{1}{2}\, \omega_\mu^{\ ab}\, M_{ab} + e^a_{\ \mu}\Pi_a = \overset{o}{D}_\mu + e^a_{\ \mu}\, \Pi_a \ . \qquad (3.31)$$

The components of the curvature tensor associated with this gauge potential may be read off from Eq. (9), the result is

$$R_{\mu\nu}^{\ \ \ ab} = \overset{o}{R}_{\mu\nu}^{\ \ \ ab} + e^a_{\ \mu}\, e^b_{\ \nu} - e^a_{\ \nu}\, e^b_{\ \mu} \ ,$$

$$R_{\mu\nu}^{\ \ \ a} = \overset{o}{D}_\mu e^a_{\ \nu} - \overset{o}{D}_\nu e^a_{\ \mu} \ . \qquad (3.32)$$

Here $\overset{o}{D}_\mu$ is the Lorentz-covariant derivative of the previous subsection, Eq. (24), and $\overset{o}{R}_{\mu\nu}{}^{ab}$ is the curvature tensor of Eq. (26).

If we could now identify the components of the gauge potential designated by $e^a{}_\mu$ with the vierbeins of the Cartan formalism, then the present framework would include all the dynamical variables of the Einstein-Cartan theory. In particular, the $R^a_{\mu\nu}$ components of the curvature in Eq. (32) could be identified with the components of the torsion tensor of Eq. (2.41).

We now want to see whether this identification is indeed possible. Let us first consider the transformation properties of the quantities $e^a{}_\mu$. With respect to Lorentz transformations these quantities just transform like ordinary Lorentz vectors. This is essentially guaranteed by the commutation relations between the associated generators Π^a and the generators of the Lorentz group given in Eqs. (29) and (30). With respect to the "inner translations" ϵ^a associated with the generators Π_a the quantities $e^a{}_\mu$ transform, on the other hand, like genuine gauge potentials, i.e.

$$\delta_\epsilon e^a{}_\mu = - D_\mu \epsilon^a . \tag{3.33}$$

As we shall see below, the quantities $e^a{}_\mu$ provide an isomorphism between the inner translations ϵ^a and the spacetime translations ϵ^μ. The inner translation ϵ^a associated with a spacetime translation ϵ^μ is just

$$\epsilon^a = e^a{}_\mu \epsilon^\mu . \tag{3.34}$$

With this result we may calculate, from Eq. (33),

$$\begin{aligned}
\delta_\epsilon e^a{}_\mu &= - \partial_\mu \epsilon^a - \mathcal{A}^B_\mu G_B \, (e^a{}_\nu \epsilon^\nu) \\
&= - \partial_\mu \epsilon^a - \epsilon^\nu f_{BC}{}^a \mathcal{A}^B_\mu \mathcal{A}^C_\nu \\
&= - \partial_\mu (e^a{}_\nu \epsilon^\nu) - \tfrac{1}{2} \epsilon^\nu (f_{bc,d}{}^a \omega_\mu{}^{bc} e^d{}_\nu + f_{b,cd}{}^a e^b{}_\mu \omega_\nu{}^{cd}) \\
&= - \epsilon^\nu (\partial_\mu e^a{}_\nu) - e^a{}_\nu (\partial_\mu \epsilon^\nu) - \epsilon^\nu \omega_\mu{}^a{}_b e^b{}_\nu + \epsilon^\nu \omega_\nu{}^a{}_b e^b{}_\mu \\
&= - \epsilon^\nu (D_\mu e^a{}_\nu - D_\nu e^a{}_\mu) - \epsilon^\nu \partial_\nu e^a{}_\mu - (\partial_\mu \epsilon^\nu) e^a{}_\nu .
\end{aligned} \tag{3.35}$$

Now we know that in Einstein-Cartan gravity the torsion vanishes. Anticipating this condition, Eq. (35) becomes:

$$\delta_\epsilon e^a{}_\mu = - \epsilon^\nu \partial_\nu e^a{}_\nu - (\partial_\mu \epsilon^\nu) e^a{}_\nu . \tag{3.36}$$

This is just the tranformation rule for an ordinary world-covariant vector field with respect to local spacetime translations, Eq. (2.4)! The general result that local coordinate transformations can be written as gauge transformations plus curvature terms, which has later found important applications in various places in the literature, was first obtained by Hehl et al.[10]

We see that if the torsion vanishes, the gauge theory based on an anti-de-Sitter group contains the same dynamical variables and *the same symmetries* as the Einstein-Cartan theory of gravity. However, in this framework the torsion corresponds to some components of the curvature tensor and, since all the components of a tensor mix under general transformations of a simple group, the condition of vanishing torsion cannot be invariant under the whole anti-de-Sitter group. It is, in fact, invariant only with respect to the Lorentz subgroup. The final, physically relevant, symmetry must thus be, not the full SO(3,2), but the subgroup SO(3,1).

The phenomenon we here encounter is called in physical terminology "spontaneous symmetry breaking", see, for example, Kibble and Stelle[11] for this point of view. I prefer to think of it, in terms of the mathematical terminology, as the *reduction of the fiber bundle* over spacetime with structure group SO(3,1). It is, in fact, proved in the mathematical literature[12] that under such a reduction the components of the gauge potential, here denoted by $e^a{}_\mu$, provide a one-to-one mapping of the four-dimensional space SO(3,2)/SO(3,1) onto the four-dimensional base manifold. Hence the matrix $e^a{}_\mu$ is indeed *invertible*, which is the main condition necessary for identifying these components of the gauge potential with the vierbeins. In general, of course, the components of the gauge potential cannot be expected to form an invertible matrix.

To summarize, the anti-de-Sitter group is the appropriate local symmetry for describing the dynamical setting of a theory of gravitation; in the end, however, only the symmetry associated with the Lorentz group is physical. This is accounted for by constructing an action which is written in terms of the anti-de-Sitter variables, but which only respects the Lorentz symmetry. This is the subject of the following subsection. It will be seen that when this is done the condition of vanishing torsion follows automatically as an Euler-Lagrange equation of the action.

In our present notation the Yang-Mills type action for a gravitional theory based on the anti-de-Sitter group is of the form

$$S \sim \int d^4x\, \epsilon^{\mu\nu\lambda\sigma} (R_{\mu\nu}{}^{ab} R_{\lambda\sigma}{}^{cd} Q_{abcd} + R_{\mu\nu}{}^{a} R_{\lambda\sigma}{}^{b} Q_{ab}) . \qquad (3.37)$$

As discussed in the previous subsection, we expect to reproduce Einstein-Cartan gravity when we make the *asymmetric* choice:

$$Q_{abcd} = \epsilon_{abcd} ; \quad Q_{ab} = 0, \qquad (3.38)$$

$$S_{adS} = \frac{1}{16} \int d^4x\, \epsilon^{\mu\nu\lambda\sigma} \epsilon_{abcd} R_{\mu\nu}{}^{ab} R_{\lambda\sigma}{}^{cd} . \qquad (3.39)$$

The invariance of this expression with respect to Lorentz transformations is obvious, its variation with respect to inner translations is given by Eq. (15):

$$\delta_\epsilon S_{adS} = \frac{1}{8} \int d^4x\, \epsilon^{\mu\nu\lambda\sigma} \epsilon_{abcd} f^{ab}_{e,g} \epsilon^{e} R_{\mu\nu}{}^{g} R_{\lambda\sigma}{}^{cd} = 0 \qquad (3.40)$$

if $R^a_{\mu\nu} = 0$. Since in this case the inner translations correspond to local spacetime translations, the action of Eq. (39), despite its not possessing the full anti-de-Sitter symmetry, is nevertheless invariant with respect to all the symmetries of the Einstein-Cartan theory.

A useful decomposition of the action is obtained by inserting the decomposition of the components of the curvature tensor as given in Eq. (32). This yields:

$$S_{adS} = \frac{1}{16} \int d^4x\, \epsilon^{\mu\nu\lambda\sigma} \epsilon_{abcd} (\mathring{R}_{\mu\nu}{}^{ab} + f_{l,f}^{ab} e^{l}{}_{\mu} e^{f}{}_{\nu}) (\mathring{R}_{\lambda\sigma}{}^{cd} + f_{h,k}^{cd} e^{h}{}_{\lambda} e^{k}{}_{\sigma})$$

$$= S_o + S_E + S_C . \qquad (3.41)$$

Here

$$S_o = \frac{1}{16} \int d^4x\, \epsilon^{\mu\nu\lambda\sigma} \epsilon_{abcd} \mathring{R}_{\mu\nu}{}^{ab} \mathring{R}_{\lambda\sigma}{}^{cd} \qquad (3.42)$$

is a topological invariant, since ϵ_{abcd} is an invariant tensor of the Lorentz group and $\mathring{R}^{ab}_{\mu\nu}$ the curvature associated with the gauge potential of this same group. Hence it makes no contribution to the equations of motion.

$$S_E = \frac{1}{8} \int d^4x\, \epsilon^{\mu\nu\lambda\sigma} \epsilon_{abcd} f_{h,g}^{cd} \mathring{R}_{\mu\nu}{}^{ab} e^{h}{}_{\lambda} e^{g}{}_{\sigma}$$

$$= \frac{1}{4} \int d^4x\, \epsilon^{\mu\nu\lambda\sigma} \epsilon_{abcd} \mathring{R}_{\mu\nu}{}^{ab} e^{c}{}_{\lambda} e^{d}{}_{\sigma}$$

$$= -\frac{1}{2} \int d^4x\, e\, \mathring{R}_{\mu\nu}{}^{ab} (e_a{}^{\mu} e_b{}^{\nu} - e_b{}^{\mu} e_a{}^{\nu}) = - \int d^4x\, e\, \mathring{R} \qquad (3.43)$$

is just the Einstein-Hilbert action in Cartan's notation, Eq. (2.57).

$$S_C = \frac{1}{16} \int d^4x\, \epsilon^{\mu\nu\lambda\sigma}\, \epsilon_{abcd}\, f^{ab}_{l,f}\, f^{cd}_{h,g}\, e^{l}_{\ \mu}\, e^{f}_{\ \nu}\, e^{h}_{\ \lambda}\, e^{g}_{\ s}$$
$$= \frac{1}{4} \int d^4x\, \epsilon^{\mu\nu\lambda\sigma}\, \epsilon_{abcd}\, e^{a}_{\ \mu}\, e^{b}_{\ \nu}\, e^{c}_{\ \lambda}\, e^{d}_{\ \sigma}$$
$$= \int d^4x\, e\, \epsilon^{abcd}\, \epsilon_{abcd} = -4!\ \frac{1}{16} \int d^4x\, e \qquad (3.44)$$

just corresponds to a cosmological constant in the Lagrangian.

As in Section 2, the Euler-Lagrange equation which results from varying the action with respect to the spin connection $\omega_{\mu ab}$ just gives the condition for vanishing torsion. The Euler-Lagrange equation which results from varying the action with respect to $e^{a}_{\ \mu}$ is

$$\overset{\circ}{R}_a^{\ \mu} - \tfrac{1}{2}\, e_a^{\ \mu}\, \overset{\circ}{R} - 3\, e\, e_a^{\ \mu} = 0\ . \qquad (3.45)$$

This is the Einstein field equation with a cosmological term. Upon contraction of the anti-de-Sitter group to the Poincaré group (see Sec. 1) we find

$$S_E \rightarrow S_E\ /\ L^2\ ; \qquad S_C \rightarrow S_C\ /\ L^4$$

and in the limit $L \rightarrow \infty$ the cosmological term drops out.

Supergravity as a Gauge Theory

The Super-AdS Group as a Local Symmetry

Our aim in the following is to demonstrate that a straightforward extension of the mathematical structures of the previous subsections from Lie algebras to Lie superalgebras leads almost automatically to an extension of Einstein gravity to Supergravity. As a first step in this direction we extend the AdS Lie algebra so(3,2) to the Lie superalgebra osp(1/4). The corresponding commutators have already been calculated in Section 1; they are exhibited in Eq. (1.59). For our present purposes it is convenient to express the results in terms of the supersymmetry generators $\bar{S}_\alpha = S^\beta C_{\beta\alpha}$. The commutators of these objects with the AdS generators and among themselves are:

$$[\bar{S}_\alpha, M_{ab}] = -(\sigma^T_{ab})_\alpha{}^\beta \bar{S}_\beta ,$$

$$[\bar{S}_\alpha, \Pi_a] = -\tfrac{1}{2}(\gamma^T_a)_\alpha{}^\beta \bar{S}_\beta ,$$

$$\{\bar{S}_\alpha, \bar{S}_\beta\} = -(C\gamma^a)_{\alpha\beta}\Pi_a + (C\sigma^{ab})_{\alpha\beta}M_{ab} . \tag{3.46}$$

From here we can read off the structure constants; they are:

$$f_{\alpha,ab}{}^\beta = -(\sigma^T_{ab})_\alpha{}^\beta ; \quad f_{\alpha,a}{}^\beta = -\tfrac{1}{2}(\gamma^T_a)_\alpha{}^\beta ,$$

$$f_{\alpha,\beta}{}^a = -\tfrac{1}{2}(C\gamma^a)_{\alpha\beta} ; \quad f_{\alpha,\beta}^{ab} = (C\sigma^{ab})_{\alpha\beta} . \tag{3.47}$$

We denote the gauge field associated with the supersymmetry generator (*the gravitino*) by ψ^α_μ ; the covariant derivative may then be written as:

$$D_\mu = \partial_\mu + \tfrac{1}{2}\omega^{ab}_\mu M_{ab} + e^a_\mu \Pi_a + \psi^\alpha_\mu \bar{S}_\alpha . \tag{3.48}$$

We shall later need the action of the covariant derivative of the Lorentz group on the gravitino field; recalling Eq. (25) this is given by

$$\overset{o}{D}_\mu \psi^\alpha_\nu = \partial_\mu \psi^\alpha_\nu + \tfrac{1}{2}\omega^{ab}_\mu (\sigma_{ab})^\alpha{}_\beta \psi^\beta_\nu . \tag{3.49}$$

Having the structure constants at our disposal we may insert them into Eq. (9) to read off the components of the curvature tensor; they are:

$$R^{ab}_\mu = \overset{o}{R}{}^{ab}_{\mu\nu} + e^a_\mu e^b_\nu - e^a_\nu e^b_\mu + \bar{\psi}_\mu \sigma^{ab} \psi_\nu ,$$

$$R^a_{\mu\nu} = \overset{o}{D}_\mu e^a_\nu - \overset{o}{D}_\nu e^a_\mu - \tfrac{1}{2}\bar{\psi}_\mu \gamma^a \psi_\nu ,$$

$$R^\alpha_{\mu\nu} = \overset{o}{D}_\mu \psi^\alpha_\nu - \overset{o}{D}_\nu \psi^\alpha_\mu + \tfrac{1}{2}e^a_\mu(\gamma_a \psi_\nu)^\alpha - \tfrac{1}{2}e^a_\nu(\gamma_a \psi^\alpha_\mu) . \tag{3.50}$$

Here $\overset{o}{D}_\mu$ and $\overset{o}{R}_{\mu\nu}$ are the covariant derivative and curvature tensor relevent to the local Lorentz group of Eqs. (25) and (27). In the next subsection we shall see that, as in the case of ordinary gravity, the dynamics of the theory leads to the vanishing of the curvature components $R^a_{\mu\nu}$. Whereas in the previous case this signalized the absense of torsion, here the equation

$$R^a_{\mu\nu} = T^a_{\mu\nu} - \tfrac{1}{2}\bar{\psi}_\mu \gamma^a \psi_\nu = 0 \tag{3.51}$$

is interpreted to mean that *spin generates torsion*.

In the AdS approach to ordinary gravity the anti-de-Sitter group was reduced to its Lorentz subgroup. In the present case we must reduce the supergroup osp(1/4) with respect to the super-Lorentz subgroup, which is generated by the osp(1/4) generators with the inner translations deleted.

The covariant derivative of the super-Lorenz group is

$$\hat{D}_\mu = \overset{o}{D}_\mu + \psi^{\bar{\alpha}}_\mu S_\alpha . \tag{3.52}$$

The components of the curvature tensor are

$$\hat{R}^{ab}_{\mu\nu} = \overset{o}{R}{}^{ab}_{\mu\nu} + f^{ab}_{\alpha,\beta}\psi^\alpha_\mu\psi^\beta_\nu ,$$

$$\hat{R}^{\alpha}_{\mu\nu} = \overset{o}{D}_\mu\psi^\alpha_\nu - \overset{o}{D}_\nu\psi^\alpha_\mu . \tag{3.53}$$

Now consider the Yang-Mills type action associated with this group:

$$\hat{S}_0 = \int d^4x\, \epsilon^{\mu\nu\lambda\sigma}\, Q_{AB}\, \hat{R}^A_{\mu\nu}\, \hat{R}^B_{\lambda\sigma} , \tag{3.54}$$

with

$$Q_{abcd} = \epsilon_{abcd} , \qquad Q_{\alpha\beta} = -i(C\gamma_5)_{\alpha\beta} . \tag{3.55}$$

The coefficient (-i) in $Q_{\alpha\beta}$ is chosen in such a way as to ensure that the variation of $\hat{S}_0$ with respect to an arbitrary variation of the gauge potential vanishes; i.e. $\hat{S}_0$ is a topological invariant. This may be easily checked by using the general formula for such variations given before.

The complete *Super AdS Action* is

$$S = \tfrac{1}{4}\int d^4x\, \epsilon^{\mu\nu\lambda\sigma}\, [\tfrac{1}{4}R^{ab}_{\mu\nu}R^{cd}_{\lambda\sigma}\epsilon_{abcd} - iR^\alpha_{\mu\nu}R^\beta_{\lambda\sigma}(C\gamma_5)_{\alpha\beta}] . \tag{3.56}$$

Here $R^{ab}_{\mu\nu}$ and $R^{\alpha}_{\mu\nu}$ are given, of course, by the general formulas:

$$R^{ab}_{\mu\nu} = \hat{R}^{ab}_{\mu\nu} + f^{ab}_{c,d}e^c_\mu e^d_\nu ,$$

$$R^\alpha_{\mu\nu} = \hat{R}^\alpha_{\mu\nu} + f^\alpha_{a,\beta}(e^a_\mu\psi^\beta_\nu - e^a_\nu\psi^\beta_\mu) . \tag{3.57}$$

As in the case of ordinary gravity, these equations suggest the following decomposition of the action:

$$S = \hat{S}_0 + S_{sg} + S_c . \tag{3.58}$$

Here $\hat{S}_0$ is, as we have seen, a topological invariant. As such it does not contribute to the equations of motion. S_{sg} is the supergravity action proper and S_c contains the cosmological constant term.

Working out the supergravity term, as in Eq. (43), we find

$$S_{sg} = - \int d^4x \; [e\overset{o}{R} + i\epsilon^{\mu\nu\lambda\sigma}\bar{\psi}_\sigma\gamma_5\gamma_\lambda(\overset{o}{D}_\mu\psi_\nu) + e\bar{\psi}_\mu\sigma^{\mu\nu}\psi_\nu] \; . \tag{3.59}$$

The first term in this expression is, of course, the familiar Einstein-Hilbert action for pure gravity, which contains the kinetic energy term for the gravitons. The second term is the so-called Rarita-Schwinger action, which contains the kinetic energy term for spin 3/2 particles, here minimally coupled to the gravitons. Note that this term, which was postulated so long ago by Rarita and Schwinger[13] on the basis of general invariance principles, comes out of the present calculation in a completely natural fashion. The last term in the above expression might be called a "gravitino mass term", although the concept of mass in the curved AdS space must be treated with some care (see Ref. [14]).

The last term in the AdS action works out to be

$$S_c = -\int d^4x \; [3e + e\bar{\psi}_\mu\sigma^{\mu\nu}\psi_\nu] \tag{3.60}$$

and contains, besides a further contribution to the gravitino mass, a cosmological constant.

Upon contraction, the cosmological constant term and the gravitino mass term fall out, and we are left with the pure *N=1 supergravity action in four dimensions:*

$$S_{sg} = -\int d^4x \; [e\overset{o}{R} + i\epsilon^{\mu\nu\lambda\sigma}\bar{\psi}_\sigma\gamma_5\gamma_\lambda(\overset{o}{D}_\mu\psi_\nu)] \; . \tag{3.61}$$

In the present treatment the invariance of this action with respect to local supersymmetry transformations is evident by construction. This is to be contrasted to the laborious calculations necessary to establish this invariance in the conventional approach.

ACKNOWLEDGEMENT

I wish to thank Prof. F. W. Hehl for providing me with some of the relevant literature on this subject.

REFERENCES

1. A. Trautman, in "General Relativity and Gravitation", A. Held, ed., Plenum, New York and London (1980).
2. S. W. MacDowell and F. Mansouri, Phys. Rev. Lett. 38: 739 (1977).
3. P. K. Townsend and P. van Nieuwenhuizen, Phys. Lett. 67B: 439 (1977).
4. R. Haag, J. T. Łopusański and M. F. Sohnius, Nucl. Phys. B88: 237 (1975).
5. S. Coleman and J. Mandula, Phys. Rev. 159: 1251 (1967).
6. M. F. Sohnius, Phys. Rep. 128: 39 (1985).
7. P. van Nieuwenhuizen, Phys. Rep. 68: 189 (1981).
8. M Scheunert, Lecture Notes in Mathematics, Springer Tracts 716, Springer, Berlin (1979).
9. T. Eguchi, P. B. Gilkey and A. J. Hanson, Phys. Rep. 66: 214 (1980).
10. F. W. Hehl, P. van der Heyde, G. D. Kerlick and J. M. Nester, Rev. Mod. Phys. 48: 393 (1976).
11. T. W. Kibble and K. S. Stelle, in "Progress in Quantum Field Theory", H. Ezawa and S. Kamefuchi, eds., North Holland, Amsterdam (1986).
12. S. Kobayashi and K. Nomizu, "Foundations of Differential Geometry", Vol. I, J. Wiley and Sons, New York (1963); see also F. Müller-Hoissen, Dissertation, Göttingen (1973) and references therein.
13. W. Rarita and J. Schwinger, Phys. Rev. 60: 61 (1941).
14. D. Z. Freedman and B. de Witt, in "Supersymmetry", K. Dietz, R. Flume, G. v. Gehlen and V. Rittenberg, eds., Plenum, New York and London (1985).

STRINGS AND SUPERSTRINGS

H. Nicolai

Institut für theoretische Physik der Universität Karlsruhe
7500 Karlsruhe 1, Fed. Rep. Germany

ABSTRACT

After a brief review of the historical development, the basic features of bosonic strings and their quantization are presented. Adding in fermionic degrees of freedeom in various ways leads to the different superstring models, including the phenomenologically promising heterotic string. String interactions are discussed, in a more pictorial than mathematical manner. Recent developments and the future outlook conclude this article.

1. INTRODUCTION

String theories have had a varied and curious history. In 1968 Veneziano[1] wrote down a four-particle scattering amplitude that embodied many of the properties that physicists expected a future theory of hadronic interactions to possess. It did not take long until it was realized that the underlying dynamics, from which the Veneziano formula could be derived, was that of a relativistic string (i.e. an extended object) rather than of a pointlike particle[2,3]. A short period of intense investigation ensued, and, at that time, several review articles were written (see reference list). However, the further development came gradually to a halt with the advent of QCD as the (probably) correct theory of strong interactions[4]. Moreover, it had become clear in 1972 that string theories could only be consistently quantized in 26 dimensions and predicted the existence of a tachyon[5,6]. At the time, this result was perhaps even more puzzling than all the failed attempts to fit the predictions of string theory with hadronic phenomena. The existence of a "critical dimension" appeared as a rather unexpected phenomenon, and it was certainly not clear why a theory that at least shared some qualitative features with hadronic physics should only work in 26 dimensions. Although one could have hoped that some kind of string theory can be derived from QCD - after all, mesons can be thought of as pairs of quarks and antiquarks bound together by a gluonic "string" - subsequent attempts in this direction have been largely unsuccessful. It appears now that a "stringy" description of hadronic physics will at best be an effective theory rather than a fundamental one.

When this point of view was adopted by the large majority of the physicists, the subject went into a period of decline, and only a few hardy people continued to work on string theories. In 1974, Scherk and Schwarz[7] suggested to radically alter the interpretation of string theories: after changing the fundamental energy scale of the theory from 200 MeV (i.e. hadronic physics) to 10^{19} GeV, by almost twenty orders of magnitude, string theories should be viewed as providing a fundamental description of all physics rather than just hadronic physics. Their rationale for this proposal was the unavoidable existence of a massless spin-2 particle in the closed string: such a particle was an embarrassment as long as one was dealing with strong interactions, but neatly fitted the properties of the graviton. However, appealing as it was, this suggestion did not attract as much attention as it would have deserved; the tachyon was still there, and the critical dimension persisted to be d=26. Also, the model did not contain any fermions.

At about the same time, supersymmetry[8] and supergravity[9,10] were formulated, and many theorists concentrated on these theories in the hope that they might provide a framework for the unification of all interactions. Especially the maximally extended N = 8 supergravity theory in 4 dimensions[11,12] seemed promising. Since supersymmetric theories tend to be less divergent than nonsupersymmetric ones[8], one of the hopes was that the local N = 8 supersymmetry of N = 8 supergravity could cure the nonrenormalizable infinities of quantum gravity. This hope was thwarted when the existence of N = 8 supersymmetric counterterms at higher orders was demonstrated (to be sure, N = 8 supergravity is suspected but has not been proven guilty of possessing higher-order divergences). Moreover, all attempts to relate N=8 supergravity to known physics failed. On the string side, it was realized in 1976 that a supersymmetric string in ten spacetime dimensions could be manufactured out of the two sectors of the spinning string[13,14] by a suitable truncation[15]. This was the birth of superstring theory.

Superstring theories are superior to the old bosonic string in several respects. The troublesome tachyon is removed from the spectrum by the truncation introduced in Ref. [15], and the critical dimension is lowered to D = 10. Moreover, the closed superstring theories contain D=10 supergravity (with or without matter), and one could therefore hope that the diseases of supergravity theory can be cured by "embedding" them in superstrings. Since superstrings contain infinitely many states, the problem of cancelling the divergences of quantum supergravity could now be reexamined in a completely new setting.

However, the formulation of Ref. [15] was not suitable to investigate these aspects; in particular, the explicit form of the supersymmetry operators relating bosonic and fermionic states was not known. For this reason, Green and Schwarz[16] developed in 1981 a "new formalism", in which the supersymmetry was explicit, and which was more appropriate to study the properties of superstrings than the "old formalism". Soon after, they were able to show that certain superstring theories (the "type II theories") were one-loop finite[17]. Since, in contrast to point-field theories, string theories are probably finite to all orders if they are one-loop finite, these results fuelled hopes that superstring theories might provide the framework for a finite theory of quantum gravity.

A dramatic increase of interest in the subject took place in 1984. One of the major problems had been to find a unified theory that predicted the left-right asymmetry of present day physics. Such a theory must inevitably contain chiral fermions. On the other hand, any theory with chiral fermions is likely to be plagued by anomalies, that is quantum mechanical breakdowns of classical conservation laws. Only in some special cases do these anomalies cancel (for example, they cancel within a standard generation of quarks and leptons), and in higher dimensions it becomes more and more difficult to achieve the requisite cancellations.

A first step had been taken in a paper of Alvarez-Gaumé and Witten in 1983[18], where the cancellation of all anomalies was demonstrated for the (chiral) type IIB theory, but this theory appeared to have no good phenomenological prospects. However, shortly thereafter it was shown[19] that, for the so-called type I theories, the requirement of anomaly cancellation in ten dimensions singles out two groups, namely SO(32) and $E_8 \times E_8$ (the cancellation of anomalies for $E_8 \times E_8$ was actually first pointed out by Thierry-Mieg[20]). Although a string theory with SO(32) symmetry existed, no string theory with $E_8 \times E_8$ symmetry was known. However, soon after this discovery, a new type of string theory was discovered - the "heterotic string", which is a hybrid of the old D = 26 bosonic string and the D = 10 superstring[21]. With this construction, it became possible to realize both SO(32) and $E_8 \times E_8$. Subsequently, the compactification of ten-dimensional supergravity coupled to $E_8 \times E_8$ matter was studied, with the result that several generations of chiral fermions could be obtained[22]. It was the first time that a theory formulated with the ambitious aim of unifying all fundamental interactions led to low energy "predictions" which were not in immediate conflict with known physics. This fact, and the hope that the theory will eventually yield unique predictions for low energy physics, have sustained the enthusiasm of many theorists ever since. Many physicists regard the heterotic $E_8 \times E_8$ string as the most promising candidate for the ultimate unification of physics. In the words of Gross et al.[21] "Although much work remains to be done there seem to be no insuperable obstacles to deriving all known physics from the $E_8 \times E_8$ heterotic string".

However, even if this optimistic assessment turns to be erroneous, there are further reasons to believe in the relevance of string theories. These theories offer much better prospects for curing the ultraviolet divergences of quantum gravity than any known point field theory. The main reason for this is the "explosion of symmetry" that takes place in string theories, which may be traced back to the special properties of the conformal group in two dimensions. This is the group of coordinate transformations that leaves distances and angles invariant up to factors (in more physical terms, conformal transformations leave the lightcone invariant). In two dimensions, all analytic functions $f(z)$ have this property, because their derivatives $f'(z)$ do not depend on the direction (in contrast, the conformal transformations form the *finitely* generated Lie-group $SO(D,2)$ in D dimensions for $D>2$). Analytic mappings are generated by the operators

$$L_m \equiv -z^{m+1} \frac{d}{dz} , \tag{1.1}$$

which satisfy the commutation relations

$$[L_m, L_n] = (m-n) L_{m+n} . \tag{1.2}$$

When reinterpreted in the framework of string theories, each of the operators L_{-m} ($m \geq 1$) gives rise to an ordinary gauge invariance in the embedding spacetime, and this "explains" why a string theory has "infinitely more" symmetry than ordinary point field theories (at this point, the reader must accept this assertion on faith; it is by no means obvious how the two-dimensional symmetries generated by the operators L_{-m} are transmuted into higher dimensional symmetries. See, however, section 6). An important result of analytic function theory is the Riemann mapping theorem (see any standard textbook on complex function theory), which states that for any two connected regions G, G' of the complex plane there exists an analytic function $f(z)$ that maps one onto the other, i.e. $f(G) = G'$. Translated into string theory this means that the physics of string theories is independent of the shape of the "world-sheet", since this shape may be conformally deformed in an arbitrary manner - like an infinitely stretchable rubber surface, as it were. Thus the physics only depends on the topology of the "world-sheet", and not on its metrical properties. To make this somewhat intuitive description mathematically precise requires a great deal of advanced mathematics.

There is also a more physical way to understand why string theories can help with the problem of quantum gravity, and this is by analogy with the theory of weak interactions. When theorists first tried to describe the decay of the neutron they did so by use of a Fermi-type Lagrangian where the interaction takes place at a point. It was later found that this theory is not renormalizable, i.e. it gives rise to irremoveable infinities at higher loop order. Nowadays, we know how to cure the problem: at sufficiently high energies (i.e. about 100 GeV), the pointlike vertex is dissolved and the weak force is mediated by a heavy boson, see Fig. 1.

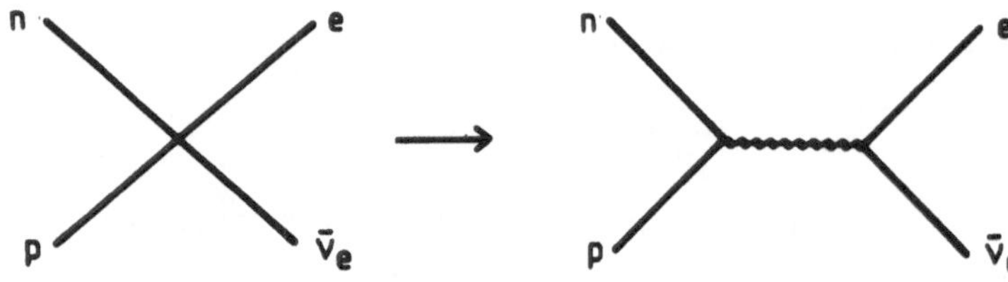

Fig. 1. A famous example of how a vertex can be dissolved.

The four-point vertex is thus replaced by a three-point vertex at high energies, and this is essentially what makes the new theory renormalizable and predictive beyond the tree aproximation.

In string theories, it is conjectured that a similar effect takes place. In Einstein's theory of gravity, one obtains n-point vertices of arbitrarily high order when expanding the action $\sqrt{-g}\, R$. These give rise to even more severe infinities than the old Fermi theory of weak interactions: the number of infinities increases with each order in perturbation theory. Contrariwise, in string theories, these vertices are dissolved just as in the above example, by the exchange of the massive string excitations (with masses quantized in units of 10^{19} GeV). In contrast to Fermi theory one has now infinitely many particles of arbitrarily high mass and "spin" to mediate these forces. In order to cope with this infinity, one needs a unifying principle, and this principle is provided by string theory.

However, there is now a much harder conceptual problem. Dissolving the gravitational vertices involves in some sense dissolving the very notion of space-time itself, because the "composite" vertices of Einstein's theory are themselves derived from an action which is based on geeometrical considerations of the structure of space-time. In string theories, this structure is replaced by something more fundamental, but it is unknown what the new principle could be. We are still accustomed to thinking of strings as moving in a flat spacetime background, but it is clear that this picture can only serve as a "crutch" towards obtaining a better understanding. In particular, we must eventually abandon the conventional notion of space-time, which should be outcome rather than input in a complete formulation of string theories.

Here we shall explain some of the basic features of string theories to the non-expert reader. More emphasis will be placed on points of principle, and the choice of presentation will reflect the author's bias as to what might survive of string theories even if the attempts to relate the currently most popular model to known physics should fail. In any case, it is hoped that this article will convey some of the excitement these models have created among theorists. The organization of this article is as follows. In section 2, we treat the classical string in analogy with the relativistic point particle. The quantization of (bosonic) string theories, the emergence of the critical dimension and the spectrum will be discussed in section 3. Section 4 is intended as an elementary introduction to superstring theory, while section 5 is devoted to a more pictorial than mathematical description of strings in interaction. The last section is meant to be an "appetizer" for those readers who want to continue with their studies of strings: it contains a short discussion of some of the topics that are currently under investigation by those on the forefront of research. Finally, we have included a list of some relevant references, without aiming at completeness (which would be impossible anyhow); the interested reader is invited to have a look at some recent issues of the relevant journals for an "entrée" into the most recent literature.

The classical theory of the relativistic string can be developed in almost complete analogy with the classical theory of a relativistic point particle moving through space-time. The world-line of such a particle is given by a function $x^{\mu} = x^{\mu}(\tau)$, see Fig.2 below.

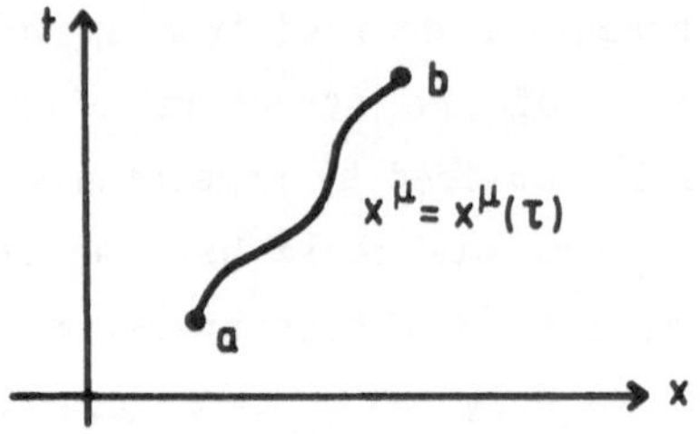

Fig.2. The world-line of a relativistic point particle

We will leave the dimension of the space-time, in which the motion takes place, arbitrary, and also assume the embedding space-time $\mathbb{R}^n$ to be flat. To describe the dynamics of this particle and to determine its trajectory we need an action, and the simplest choice is simply the "length" of the trajectory (this is not the Euclidian length but rather a Minkowskian length in a space of signature (-+...+)). Thus we put

$$S = \text{"length"} = -m \int_a^b \sqrt{-\dot{x}^{\mu}\dot{x}_{\mu}}\; d\tau \,, \tag{2.1}$$

where the parameter m has the dimension mass = $[\mathrm{cm}]^{-1}$, and is needed to render the action dimensionless. A most important feature of (2.1) is its invariance under reparametrizations $\tau \rightarrow \tau'(\tau)$: the physics should not depend on how the trajectory is parametrized. The positivity of $-\dot{x}^2$ in the integral is equivalent to the requirement that a particle should not move faster than at the speed of light. With (2.1) we can calculate the canonical momenta p^{μ} which are associated with the coordinates $x^{\mu}(\tau)$; we obtain

$$p^{\mu} = \frac{\partial}{\partial \dot{x}_{\mu}} \left(-m \sqrt{-\dot{x}^2} \right) = m\,(-\dot{x}^2)^{-\frac{1}{2}}\, \dot{x}^{\mu} \,. \tag{2.2}$$

These momenta are not independent, but rather obey the constraint

$$p^2 + m^2 = 0 \,, \tag{2.3}$$

as one can straightforwardly verify from (2.2). This is the well-known "dispersion relation" of a relativistic point particle. The fact that the canonical momenta are constrained complicates the Hamiltonian formalism somewhat; it is a reflection of the invariance of (2.1) under reparametrizations (the Hamiltonian formalism with constraints has been developed by Dirac[23], see also N. Falck's lectures in this volume). In the quantum theory, (2.3) becomes a constraint on the physical states and, after the replacement $p^{\mu} \rightarrow i\partial/\partial x^{\mu}$, it is nothing but the Klein-Gordon equation. This method of deriving the Klein-Gordon equation from the classical theory of a relativistic point-particle may seem unusual, but it is possible to derive and develop all of quantum field theory on this basis. Of course, we are more used to a formulation in terms of second quantization, i.e. involving quantum fields, but the merits of the above method have been recognized only recently in connection with string theories, where the analog of the second quantized formulation is still being developed.

The classical physics of strings can be described in an analogous fashion. The basic object is now a string, that is an extended object, rather than a point. During its motion, it sweeps out a "world-sheet" rather than a world-line. To parametrize this world-sheet we need an extra parameter σ, which conventionally is taken to lie in the interval $[0,\pi]$. Hence the motion is completely described by the functions $x^{\mu} = x^{\mu}(\sigma,\tau)$. Strings come in two varieties, namely "open" and "closed", which are distinguished through their boundary conditions.

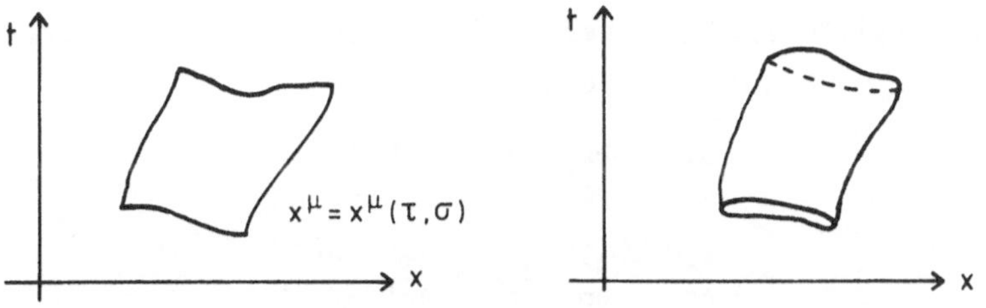

Fig. 3. Open and closed strings

Again, the dimension of the embedding space $\mathbb{R}^D$ is left arbitrary for the moment; it will, however, be restricted in the quantized theory, in contrast to the case of the point particle. To describe the dynamics of the string, we again make the simplest choice, namely[2]

$$S = \text{"surface area"} = -\frac{1}{2\pi\alpha'}\int\sqrt{(\dot{x}x')^2 - \dot{x}^2x'^2}\,d\sigma\,d\tau\ , \qquad (2.4)$$

where $\dot{x}^{\mu} \equiv \partial x^{\mu}/\partial\tau$, $x^{\mu'} \equiv \partial x^{\mu}/\partial\sigma$, and the parameter α' has dimension $[\text{mass}]^{-2} = [\text{cm}]^{+2}$, the quantity $1/2\pi\alpha'$ (of dimension $[\text{cm}]^{+2}$) is called the string tension. The expression (2.4) is invariant under reparametrizations $\sigma \to \sigma'(\sigma,\tau)$, $\tau \to \tau'(\sigma,\tau)$; the positivity of the integrand means that the string moves no faster than at the speed of light at any of its points. The reparametrization invariance of (2.4) plays an even more important role than in the case of a point particle. In particular, it allows us to choose an "orthonormal" gauge, where

$$\dot{x}^2 + x'^2 = 0 \, , \quad \dot{x}.x' = 0 \, . \tag{2.5}$$

(2.5) simply means the following: if we cover the world sheet by a mesh of lines, the lines of constant σ and τ will intersect at right angles everywhere, see Fig. 4.

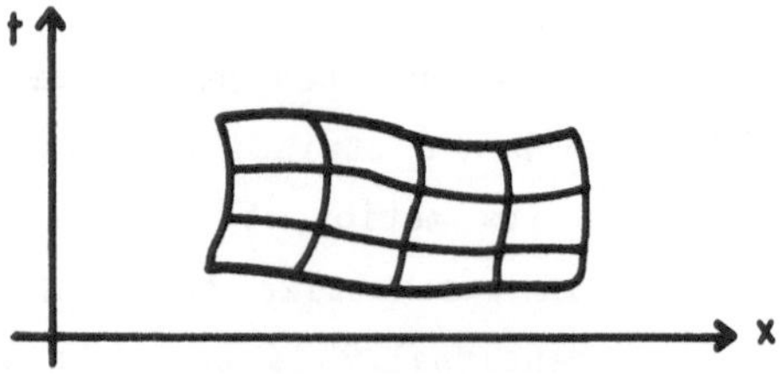

Fig.4. The orthonormal gauge

At this point, one can already see why two-dimensional surfaces (and therefore strings) are so special as opposed to higher dimensional objects (corresponding to membranes, etc.). The gauge choice (2.5) does not fix the gauge completely, but there is a huge residual invariance which preserves (2.5)! To find it, we use

$$\frac{\partial x^\mu}{\partial \sigma} = \frac{\partial x^\mu}{\partial \sigma'}\frac{\partial \sigma'}{\partial \sigma} + \frac{\partial x^\mu}{\partial \tau'}\frac{\partial \tau'}{\partial \sigma}$$

$$\frac{\partial x^\mu}{\partial \tau} = \frac{\partial x^\mu}{\partial \sigma'}\frac{\partial \sigma'}{\partial \tau} + \frac{\partial x^\mu}{\partial \tau'}\frac{\partial \tau'}{\partial \tau} \, , \tag{2.6}$$

and require that (2.5) be valid also in terms of the new coordinates (σ',τ'). After a little calculation, one finds that this implies

$$\frac{\partial \sigma'}{\partial \tau} = \frac{\partial \tau'}{\partial \sigma} \, , \quad \frac{\partial \sigma'}{\partial \sigma} = \frac{\partial \tau'}{\partial \tau} \, . \tag{2.7}$$

Introducing complex variables $z \equiv \sigma + i\tau$, $z' \equiv \sigma'(\sigma,\tau) + i\tau'(\sigma,\tau) = f(z)$, we see that (2.7) is equivalent to $(\bar{z} \equiv \sigma - i\tau)$

$$\frac{\partial f}{\partial \bar{z}} = \frac{1}{2}\left[\frac{\partial}{\partial \sigma} + i\,\frac{\partial}{\partial \tau}\right] f(\sigma,\tau) = 0 \, . \tag{2.8}$$

Therefore, (2.7) is equivalent to the well-known Cauchy-Riemann equations, telling us that f(z) is an analytic function. Now, the set of analytic functions constitutes a huge class of transformations of the complex plane, or, more generally, of a two-dimensional surface (possibly with handles and holes). This class is much larger than the corresponding set of conformal transformations in dimensions higher than two, simply because there is no analog of analytic function theory in higher dimensions. There the conformal transformations are given by a *finite* set of functions. In more technical terms, for D>2 the conformal transformations (which preserve angles and distances) form the Lie-group SO(D,2), which is generated by finitely many transformations, whereas for D=2 the group of analytic transformations has infinitely many generators, see (1.1).

Application of the canonical Hamiltonian formalism to the Lagrangian that one extracts from (2.4) leads to the same difficulties as for the point particle: owing to the reparametrization invariance of (2.4) the canonical momenta are constrained (there are now infinitely many of them, one corresponding to each $x^{\mu}(\sigma)$). Rather than discuss this in detail (see however Scherk[24]), we now proceed directly to the equations of motion that follow from (2.4). To be able to drop surface terms, we impose the following boundary conditions

$$x'(0) = x'(\pi) = 0 \ , \quad \text{(open string)}$$
$$x(0) = x(\pi) \ . \quad \text{(closed string)} \tag{2.9}$$

Varying (2.4) directly leads to some rather complicated equations which can be considerably simplified by use of (2.5). In this gauge, the equation of motion of the string is nothing but the free wave equation in two dimensions, namely

$$\ddot{x}^{\mu} - x''^{\mu} = 0 \ . \tag{2.10}$$

This is now easy to solve; we get (we put $2\alpha' = 1$ from now on)

$$x^{\mu}(\sigma,\tau) = q^{\mu} + p^{\mu}\tau + i \sum_{n\neq 0} \frac{1}{n} \alpha_n^{\mu} \cos n\sigma \, e^{-in\tau} \tag{2.11}$$

for the open string, and

$$x^{\mu}(\sigma,\tau) = q^{\mu} + p^{\mu}\tau + \frac{i}{2} \sum_{n\neq 0} \frac{1}{n} \left[\alpha_n^{\mu} \, e^{-2in(\tau-\sigma)} + \bar{\alpha}_n^{\mu} \, e^{-2in(\tau+\sigma)} \right] \tag{2.12}$$

for the closed string. In both (2.11) and (2.12), the term $q^{\mu}+p^{\mu}\tau$ describes the center-of-mass motion of the string. The other modes describe its internal motions (vibrations). Note that the closed string contains twice as many modes as the open string, corresponding to the left and right moving waves on the string.

To incorporate the constraints (2.5), we note that for the open string ($\alpha_0^\mu \equiv p^\mu$)

$$\dot{x}^\mu \pm x'^\mu = \sum_{n=-\infty}^{\infty} \alpha_n^\mu e^{-in(\tau\pm\sigma)} , \tag{2.13}$$

and therefore (2.5) is equivalent to

$$(\dot{x} \pm x')^2 = 2 \sum_{m=-\infty}^{\infty} L_m e^{-im(\tau\pm\sigma)} = 0 , \tag{2.14}$$

where we have defined the Fourier modes

$$L_m = \tfrac{1}{2} \sum_{n=-\infty}^{\infty} \alpha_{m-n}^\mu \alpha_{n\mu} . \tag{2.15}$$

Thus (2.5) holds if $L_m=0$ for all m. For the closed string, there is another $\bar{L}_m$ associated with the $\bar{\alpha}$-modes. It is instructive to analyze the L_0 constraint a little further; we have

$$L_0 = \tfrac{1}{2} p^2 + \tfrac{1}{2} \sum_{n=1}^{\infty} \alpha_{-n}^\mu \alpha_{n\mu} = 0 . \tag{2.16}$$

This is the string analog of (2.3); owing to the infinitely many internal excitations the mass of the string can assume infinitely many values, such that a particle mass is associated with a particular vibrational excitation of the string. From (2.16) it might appear that $M^2=-p^2$ is not positive, due to the indefiniteness of the Minkowski metric. However, this is a gauge artifact, as we shall see below.

Instead of just imposing the constraints $L_m=0$ on the system, one may alternatively eliminate all unphysical degrees of freedom. For this purpose, one introduces light cone coordinates

$$x^\pm \equiv \frac{1}{\sqrt{2}} (x^0 \pm x^{D-1}) , \quad x^i (i=1,\ldots,D-2) \tag{2.17}$$

where x^i are the transverse coordinates. One can then show (see e.g. Scherk[24]) that the residual gauge invariance of (2.5) is entirely fixed by putting

$$x^+(\sigma,\tau) = q^+ + p^+\tau , \tag{2.18}$$

i.e. gauging to zero all the α_m^+ excitations. Substituting (2.18) into (2.5), we can solve for $x^-(\sigma,\tau)$,

$$x^-(\sigma,\tau) = q^- + p^-\tau + i \sum_{m\neq 0} \alpha_m^- \cos m\sigma \, e^{-im\tau} , \tag{2.19}$$

where the α_m^-'s are now expressed as functions of the transverse excitations

$$\alpha_m^- = \frac{1}{2p_+} \sum_{n=-\infty}^{\infty} \alpha_{m-n}^i \, \alpha_n^i \, . \tag{2.20}$$

Observe that, up to the factor p_+^{-1}, α_m^- is just like L_m in (2.15), but with the sum ranging only over transverse indices. Furthermore, putting $\alpha_m^+ = 0$ in (2.16) we see that it now reads

$$M^2 = - p^2 = 2 \sum_{n=1}^{\infty} \alpha_{-n}^i \, \alpha_n^i \, , \tag{2.16'}$$

which is manifestly positive. All physical degrees of freedom are now contained in the transverse oscillations (and the center-of-mass coordinates and momenta). To understand why this is true one must investigate the theory in more detail, but even without doing so, the analogy with electrodynamics may be helpful. There, the electromagnetic potential or photon field $A^{\mu}(x)$ has four components. In momentum space one may decompose these components into time-like, longitudinal and transverse ones, with respect to the momentum four-vector of the photon. Owing to gauge invariance, only the transverse components of the electromagnetic potetial carry physical information, while the other components are gauge degrees of freedom. In string theories, the conformal transformations take over the role of gauge invariance, in that they may be used to eliminate the time-like and longitudinal components of the string. This analogy shows that the importance of conformal invariance in string theories cannot be overemphasized.

3. QUANTIZATION, CRITICAL DIMENSION AND SPECTRUM

String theories possess some very remarkable properties which reveal themselves only after quantization. The most remarkable one is that string theories can be consistently quantized only in certain critical dimensions (recall that there were no such restrictions for the relativistic point particle, which can be quantized in any dimension). Furthermore, in the bosonic string theories, quantization forces the ground state of the string to be a tachyon - a particle of imaginary mass that travels faster than light. Obviously, neither of these properties were especially welcome to the physicists who tried to describe hadronic physics with string theories, but nowadays, with a completely changed perspective, one tends to view these features as virtues rather than as shortcomings of the theory. Moreover, in superstring theories, the tachyon disappears and the critical dimension is lowered.

The most powerful approach to quantize string theories is through functional integral methods introduced by Polyakov[25]. This method is well suited to compute higher order "radiative" corrections, but is technically demanding, and requires an intimate knowledge of advanced mathematics such as Riemann surface theory. For this reason, we shall not dwell on this topic, but rather stick to the more conventional approach (which also has its pitfalls!). In the foregoing section, it has been explained that the relativistic string behaves in many ways like an ordinary violin string. Apart from the constraint (2.5), it satisfies a free two-dimensional wave equation (2.10) which can be easily solved, see (2.11) and (2.12). The motion is described through the modes α_n^μ (or α_m^μ and $\bar{\alpha}_n^\mu$), which are just ordinary harmonic oscillators. It is therefore not surprising that, after a canonical treatment (see e.g. Scherk[24]), the quantized string is a collection of infinitely many harmonic oscillators. More precisely, the modes α_m^μ become creation operators for m<0 and annihilation operators for m>0, which are subject to the commutation relations

$$[\alpha_m^\mu, \alpha_n^\nu] = m\ \delta_{m+n,0}\ \eta^{\mu\nu} \quad , \quad (\alpha_m^\mu)^+ = \alpha_{-m}^\mu \ . \tag{3.1}$$

The center-of-mass coordinates and momenta obey the usual commutation relations

$$[q^\mu, p^\nu] = i\ \eta^{\mu\nu} \ . \tag{3.2}$$

The Fock space H of the theory is the product of the single harmonic oscillator Hilbert spaces, and consists of all states of the form

$$\Psi = \prod_r \alpha_{-n_r}^{\mu_r} \ |0,k\rangle \ . \tag{3.3}$$

The ground state $|0,k\rangle$ has momentum k, and is annihilated by all oscillators a_m^μ with m≥1. Consequently, all other expressions now become "operator-valued". For instance, (2.15) is now an operator in H. However, some care must be exercised because, unlike the classical expressions, these operators may cease to be well defined. In particular, the sum over the oscillator vacuum energies diverges like $\frac{1}{2}\sum_{n=1}^{\infty} n$, as one can easily verify by computing the vacuum expectation value of L_0. To avoid this problem, one modifies all potentially ill-defined operators by moving the annihilation operators to the right; this amounts to subtracting off all infinities. This procedure is referred to as "normal ordering", and denoted by semicolons. For instance, we have

$$L_0 = \frac{1}{2} \sum_{n=-\infty}^{\infty} : \alpha_{-n}^{\mu} \alpha_{n\mu} : = \frac{1}{2} p^2 + \sum_{n=1}^{\infty} \alpha_{-n}^{\mu} \alpha_{n\mu} , \tag{3.4}$$

such that $\langle 0|L_0|0\rangle$ is now well defined. The normal ordering leads to a very important modification in the algebra of the L_m-operators, which now reads (a derivation of the extra terms is given in Scherk[25])

$$[L_m,L_n] = (m-n)\, L_{m+n} + \frac{D}{12}\, m(m^2-1)\, \delta_{m+n,0} \,. \tag{3.5}$$

The "central term" in (3.5) may be viewed as an "anomaly": with it, the L_m's no longer generate the algebra of conformal transformations (1.2). The new contribution in (3.5) is the source of all the peculiarities that distinguish the quantized string from its classical counterpart. To restore conformal invariance, we are forced to a particular value of D and to a tachyonic ground state.

An obvious problem is already raised by the relation (3.1). Choosing the time-like excitation α_m^0 and remembering $\eta^{00}=-1$, we can easily calculate the norm of the state $\alpha_{-m}^0|0,k\rangle$ ($m>0$)

$$\langle 0,k|\, \alpha_m^0\, \alpha_{-m}^0 \,|0,k\rangle = \langle 0,k|\, [\alpha_m^0\, \alpha_{-m}^0]\, |0,k\rangle = -m < 0 \,, \tag{3.6}$$

which is negative! This result is incompatible with the usual lore of quantum mechanics, where the norm of a state is interpreted as a probability, which should be positive. We must therefore devise a method to get rid of these "negative norm states". The clue to the solution is conformal invariance, and it is analogous to the solution of a related problem in quantum electrodynamics, see also the remarks at the end of section 2. There, the time-like component of the photon leads to a negative norm state, but this state can be eliminated by a gauge transformation. One can do this either by imposing a gauge condition on physical states, in which case one has to prove that no negative norm states are left, or by going to a light cone gauge, which contains only the transverse photon components. In the second case there are evidently no negative norm states, but one must show by explicit computation that Lorentz invariance is not violated. These two ways of eliminating unphysical states have their analogs in string theory, but here the framework is much more restrictive.

Let us first discuss the method of defining the physical states by constraints. In the classical theory we have the constraint (2.5), which is equivalent to $L_m=0$ for all m. It is easy to see, however, that we cannot impose L_m to vanish on the physical states for all m. Namely, inserting (3.5) between two physical states would lead to a contradiction immediately, because of the "anomaly". Rather, as in the Gupta-Bleuler formulation of quantum electrodynamics, one must relax the condition by imposing this requirement only for "positive frequency" operators, that is

$$L_m \, |phys\rangle = 0 \quad \text{for } m \geq 1 \, . \tag{3.7}$$

For L_0, one must allow for an extra shift

$$(L_0 - \alpha(0)) \, |phys\rangle = 0 \, , \tag{3.8}$$

where the intercept $\alpha(0)$ must be determined from the consistency requirement. Because of $L_{-m} = L_m^+$, (3.7) implies

$$\langle phy | \, L_m \, |phys'\rangle = 0 \quad \text{for all } m \neq 0 \, , \tag{3.9}$$

so, in the classical limit, we recover (2.5).

The problem is now the following. By imposing (3.7) and (3.8), we single out a subspace H_{phys} of the full Hilbert space H spanned by the states (3.3). Under what conditions can one prove that H_{phys} is free of negative norm states? The answer to this question can only be given after a lengthy argument which we shall not reproduce here (see Brower[5], Goddard and Thorn[6], Goddard et al.[26]). It turns out that things work out only if

$$D = 26 \, , \quad \alpha(0) = 1 \, . \tag{3.10}$$

Assuming $\alpha(0)=1$, one can make the following plausibility argument for the emergence of the number D=26. Consider the following state for arbitrary D

$$\Psi = [\, \alpha^{\mu}_{-1} \alpha_{-1\mu} + A \, k_{\mu} \alpha^{\mu}_{-2} + B(k_{\mu}\alpha^{\mu}_{-1})^2 \,] \, |0,k\rangle \, , \tag{3.11}$$

and impose the physical state constraints (3.7) and (3.8) on Ψ. It is actually sufficient to consider only the operators

$$\begin{aligned} L_0 &= \tfrac{1}{2} p^2 + \alpha_{-1}\alpha_1 + \alpha_{-2}\alpha_2 + \dots \\ L_1 &= p \, \alpha_1 + \alpha_{-1}\alpha_2 + \dots \\ L_2 &= \tfrac{1}{2} \alpha_1^2 + p \, \alpha_2 + \dots \end{aligned} \tag{3.12}$$

as the higher mode oscillators annihilate the state (3.11), and $L_3=[L_2,L_1]$ etc. $(L_0-1)\Psi=0$ leads to $k^2=-2$, while the L_1 and L_2 constraints lead to

$$A = \frac{D-1}{5}, \quad B = \frac{D+4}{10}. \tag{3.13}$$

The norm of the state Ψ for arbitrary D is

$$\langle\Psi|\Psi\rangle = \frac{2}{25}(26-D)(D-1). \tag{3.14}$$

Thus, $\langle\Psi|\Psi\rangle < 0$ for $D > 26$, in which case H_{phys} contains negative norm states, and there is no hope of consistently quantizing the theory. For D=26, $|\Psi\rangle$ is a zero norm state which does not correspond to a physical excitation (like the state with equally many time-like and longitudinal photons in quantum electrodynamics), and a consistent quantum theory exists. For D<26, a consistent theory may exist, but would contain extra states. Although the above argument proves the inconsistency for D>26, it is, of course, not the whole story. But we hope at least that it gives the reader an idea as to where the number 26 comes from.

An alternate approach to quantization is by solving the constraints first as in (2.19), and expressing everything through the transverse oscillators α^i_m. In this case, all operators which were responsible for the occurence of negative norm states have dissapeared, and unitarity is manifest. On the other hand, manifest Lorentz invariance has been lost, and one must check explicitly whether it can be restored. After a tedious calculation one recovers the conditions (3.10), and therefore the two approaches are entirely equivalent[26].

The light-cone gauge is actually somewhat more convenient to describe the physical spectrum of string theories, as it contains no unphysical operators. Taking into account the shift by $\alpha(0)=1$, the quantum analog of the mass formula (2.16') is

$$M^2 = \sum_{n=1}^{\infty} \alpha^i_{-n}\,\alpha^i_n - 1. \tag{3.15}$$

Unlike in (2.16'), where M^2 varies continuously, M^2 has only integer values in the quantum theory. The lowest state $|0,k\rangle$ has no oscillator excitations, and therefore $M^2 = -1$. The next state is $\alpha^i_{-1}|0,k\rangle$, which has $M^2=0$. Since this state has only transverse excitations, it is like the photon. Continuing in this fashion, one obtains the following states, ordered according to increasing mass.

TABLE 1 Open String Spectrum

$M^2 = -1$	$\|0,k\rangle$	Tachyon
$M^2 = 0$	$\alpha_{-1}^{i}\|0,k\rangle$	"Photon"
$M^2 = 1$	$\alpha_{-1}^{i}\alpha_{-1}^{j}\|0,k\rangle$ $\alpha_{-2}^{i}\|0,k\rangle$	Massive "Spin-2" Excitation
$M^2 = 2$	$\alpha_{-1}^{i}\alpha_{-1}^{j}\alpha_{-1}^{k}\|0,k\rangle$ $\alpha_{-1}^{i}\alpha_{-2}^{j}\|0,k\rangle$ $\alpha_{-3}^{i}\|0,k\rangle$	

etc.

The spectrum of the closed string can be analyzed in a similar fashion. We have already mentioned that there are twice as many oscillators in this case. The condition that there should be no distinguished point on the closed string leads to an additional constraint on the physical states, namely (in the light-cone gauge)

$$\sum_{n=1}^{\infty} (\alpha_{-n}^{i} \alpha_{n}^{i} - \bar{\alpha}_{-n}^{i} \bar{\alpha}_{n}^{i}) \, |\text{phys}\rangle = 0 \, . \tag{3.16}$$

i.e. the number of unbarred and barred excitations must be the same. The mass formula for the closed string is

$$M^2 = \sum_{n=1}^{\infty} (\alpha_{-n}^{i} \alpha_{n}^{i} - \bar{\alpha}_{-n}^{i} \bar{\alpha}_{n}^{i}) - 2 \, , \tag{3.17}$$

and the lowest state is therefore again a tachyon. Because of (3.16) neither of the states $\alpha_{-1}^{i}|0,k\rangle$ or $\bar{\alpha}_{-1}^{i}|0,k\rangle$ belongs to the physical spectrum. The next state is therefore

$$\Psi^{ij} = \alpha_{-1}^{i}\bar{\alpha}_{-1}^{j} \; |0,k\rangle \, , \tag{3.18}$$

which is massless because of (3.17). One can decompose Ψ^{ij} into irreducible parts according to

$$\Psi^{ij} = \Psi_1^{(ij)} + \delta^{ij}\Psi_2 + \Psi_3{}^{[ij]} \, , \tag{3.19}$$

where $\Psi_1^{(ij)}$ is symmetric and traceless in (ij), and $\Psi_3^{[ij]}$ is antisymmetric in [ij]. A symmetric traceless two-index tensor describes a spin-2 particle (at least in 4 dimensions). It was coincidence that inspired Scherk and Schwarz[7] to make the identification

$$\Psi_1^{(ij)} = \text{"Graviton"} \tag{3.20}$$

(The state Ψ_2 is referred to as "dilaton"). One of the remarkable things about string theory is that the existence of this "graviton" is a *prediction* rather than an input: even if one starts with open strings, which contain only a "photon", the "graviton" arises as an intermediate state. One is therefore inevitably forced to include gravity in the unification; there is no consistent string theory without gravity!

Before passing on, we make two further comments on the open string spectrum of table 1 (similar remarks apply to the closed string spectrum). Although we have not explained the notion of "spin" in 26 dimensions, it is evident from the table that there is a correlation between M^2 and the "spin". In fact, a plot reveals that the states lie on so-called "Regge-trajectories", see Fig.5. below (There are also many "daughter trajectories" which are not shown).

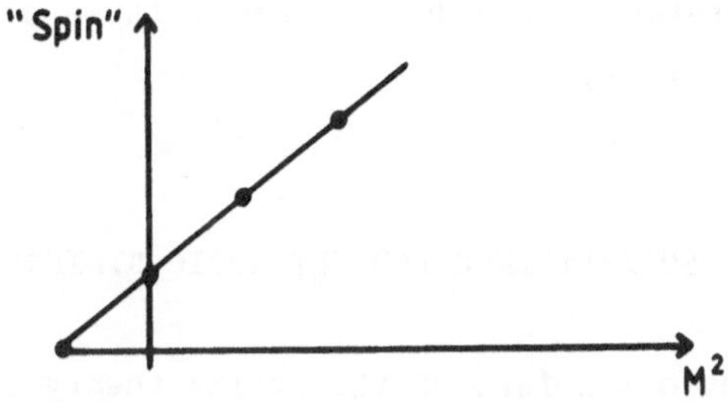

Such "Regge-trajectories" were in fact observed in the sixties in the form of mesonic and baryonic resonances. In the modern interpretation, of course, the higher excited states on these trajectories have masses of the order of 10^{19} GeV, and are therefore unobservable.

A second noteworthy feature of the string spectrum is the enormous increase in the number of states as one goes to higher and higher values of M^2. One can show that the number of states $n(M)$ at a given mass level M^2 asymptotically grows like

$$n(M) \sim \text{const.} \left[\frac{M}{M_o}\right]^{\alpha} \exp\left[\frac{M}{M_o}\right] , \tag{3.21}$$

where α depends on the dimension, and M_o is related to the fundamental scale $(\alpha')^{-\frac{1}{2}}$ of the theory (i.e. 1 GeV for hadronic physics and 10^{19} GeV for gravitational physics). Inserting (3.21) into the usual formula for the free energy (actually, this formula should also include an integral over all 26-dimensional momenta, but its omission does not affect our main conclusion)

$$F(T) = -kT \log \left[\sum_M n(M) \exp\left[-\frac{E(T)}{kT}\right] \right] \tag{3.22}$$

we see that the sum diverges at the critical temperature

$$kT_{crit} = M_0 \ . \tag{3.23}$$

This result indicates that, at this temperature, a phase transition takes place. A natural interpretation is that at $T = T_{crit}$, the string "breaks up" into its constituents, e.g. quarks and gluons, which then form a plasma for $T > T_{crit}$. In fact, in the old days, the interpretation was slightly different. It was assumed that when one tries to heat hadronic matter beyond T_{crit}, it cooled down again by "boiling off" mesons, etc. Since therefore T_{crit} is a sort of ultimate temperature, it was sometimes referred to as "hell's temperature"[27]. At present, one does not understand what happens when a gravitational string is heated beyond T_{crit} and what its constituents could be. Perhaps these are indications of a theory beyond string theory.

4. SPINNING STRINGS, SUPERSTRINGS AND HETEROTIC STRINGS

There is one obvious defect of the string theory discussed so far: it describes only bosons. In 1971, Ramond[13], and Neveu and Schwarz[14] proposed new models in an attempt to remedy this fact. Both models share the feature that, in addition to the string coordinate $x^\mu(\sigma,\tau)$, they contain its fermionic partner $\lambda^\mu(\sigma,\tau)$, whose modes satisfy anticommutation relations rather than commutation realtions. One can visualize this by attaching a fermion to every point on the world sheet. The mode expansion of the fermionic field $\lambda^\mu(\sigma,\tau)$ is quite analogous to (2.11) and (2.12), and is given by

$$\lambda^\mu(\sigma\pm\tau) = \sum_n d_n^\mu \, e^{-in(\sigma\pm\tau)} \tag{4.1}$$

in the Ramond sector, and

$$\lambda^\mu(\sigma\pm\tau) = \sum_r b_r^\mu \, e^{-in(\sigma\pm\tau)} \tag{4.2}$$

in the Neveu-Schwarz sector. To be precise, λ^μ is a two-dimensional fermion and thus should have two components. The Dirac equation in two dimensions, and the boundary conditions, imply that the upper (lower) components are given by the *same* function $f(\sigma+\tau)$ (or $f(\sigma-\tau)$), and therefore we write out only one component, see Scherk[24]. A significant difference between (4.1) and (4.2) is that the oscillators in (4.1) are integer-moded (n=....,-2,-1,0,1,2..), whereas they are half-integer moded (r=....,-3/2, -1/2, 1/2, 3/2,...) in (4.2). The anticommutation relations are

$$\{d^\mu_m, d^\nu_n\} = \delta_{m+n,0}\ \eta^{\mu\nu} ,$$

$$\{b^\mu_r, b^\nu_s\} = \delta_{r+s,0}\ \eta^{\mu\nu} . \tag{4.3}$$

For m=n=0, we obtain $\{d^\mu_0, d^\nu_0\} = \eta^{\mu\nu}$, which means that d^μ_0 is like a γ-matrix: it implies that the ground state is a fermion in space-time. On the other hand, no such peculiarity occurs in the Neveu-Schwarz sector, whose ground state is a boson. Since both d^μ_m and b^μ_r behave like vectors under space-time Lorentz transformations, the spacetime statistics is not changed by the action of the oscillators of the ground state, and therefore the Ramond spectrum contains only fermions, and the Neveu-Schwarz spectrum contains only bosons. It may seem paradoxical at first sight that, although we started with two-dimensional fermionic operators, the resulting spectrum in the embedding spacetime may consist of either bosons or fermions. However, it is well-known among the experts (Coleman[28]; Mandelstam[29]) that in two dimensions bosons and fermions are equivalent, and therefore the two-dimensional statistics implies nothing about the statistics in space-time. From the above remarks, it follows that the space-time statistics depends only on the "modedness" of the oscillators, or, more precisely, on the boundary conditions obeyed by λ^μ!

To study the spectrum of these models, we switch again to the light-cone gauge. The mass formulas for the open spinning strings are given by

$$M^2 = \sum_{n=1}^{\infty} \alpha^i_{-n}\, \alpha^i_n + \sum_{n=1}^{\infty} n\, d^i_{-n}\, d^i_n \tag{4.4}$$

in the Ramond sector, and by

$$M^2 = \sum_{n=1}^{\infty} \alpha^i_{-n}\, \alpha^i_n + \sum_{r=\frac{1}{2}}^{\infty} r\, b^i_{-r} b^i_r - \tfrac{1}{2} \tag{4.5}$$

in the Neveu-Schwarz sector. In the Ramond sector the ground state is a fermion which furthermore obeys the Dirac equation (we will not derive this here, see e.g. Scherk[24] for a more detailed discussion); this ground state has no further excitations, and is therefore massless by (4.4). The ground state in the Neveu-Schwarz sector has $M^2 = -\frac{1}{2}$, and is again a tachyon. The mass values are obtained by considerations similar to the ones that led to (3.10). In addition, the value of the critical dimension is lowered to

$$D = 10 \tag{4.6}$$

for spinning strings (again, the calculation required to prove this is very tedious and will not be reproduced here, see e.g. Green and Schwarz[16]).

The excited states are simply obtained by acting with the oscillators on the ground state. In the Neveu-Schwarz sector, this procedure leads to the following table

TABLE 2 Neveu-Schwarz Spectrum

$M^2 = -\frac{1}{2}$	$\|0,k\rangle$	Tachyon
$M^2 = 0$	$b^i_{-\frac{1}{2}}\|0,k\rangle$	"Photon"
$M^2 = +\frac{1}{2}$	$b^i_{-\frac{1}{2}}b^j_{-\frac{1}{2}}\|0,k\rangle$ $\alpha^i_{-1}\|0,k\rangle$	Massive "Spin-2" Excitation
$M^2 = +1$	$b^i_{-\frac{1}{2}}b^j_{-\frac{1}{2}}b^k_{-\frac{1}{2}}\|0,k\rangle$ $b^i_{-\frac{1}{2}}\alpha^j_{-1}\|0,k\rangle$ $b^i_{-\frac{3}{2}}\|0,k\rangle$	

etc.

In the Ramond sector we obtain

TABLE 3 Ramond Spectrum*

$M^2 = 0$	$\|0,k\rangle$	Massless Fermion
$M^2 = 1$	$\alpha^i_{-1}\|0,k\rangle$ $d^i_{-1}\|0,k\rangle$	Massive "Spin-3/2" Excitation

etc.

* $|0,k\rangle$ is a spinor wave function

It was not until 1976 that it was recognized that, by combining these two spectra in a suitable way, a supersymmetric spectrum could be obtained in ten dimensions, i.e. a spectrum containing equally many bosons and fermions at each mass-level (Gliozzi, Scherk and Olive[15]). The clue was the elimination of half the states in each sector. Obviously, we must eliminate all states of half integer M^2 in the Neveu-Schwarz sector since these cannot have fermionic partners in the Ramond sector, where M^2 assumes only integer values. This is equivalent to discarding all states created by an even number of b-oscillators; observe that the troublesome tachyon is also eliminated in this way! The truncation to states with an odd number of b-oscillators is implemented by the projection operator

$$P_{NS} = \tfrac{1}{2} \left[1 - (-1)^{\sum_{r=\frac{1}{2}}^{\infty} b^i_{-r} b^i_r} \right] . \tag{4.7}$$

Although it is not so obvious, a similar truncation is needed in the Ramond sector, which is accomplished by means of the projection operator

$$P_R = \tfrac{1}{2} \left[1 - \gamma^5 (-1)^{\sum_{n=1}^{\infty} d^i_{-n} d^i_n} \right] , \tag{4.8}$$

where γ^5 is the ten-dimensional analog of the usual γ^5 matrix. The supersymmetry of this truncated system at the massless level is easily checked: a "Majorana-Weyl-spinor" in D=10 has eight real components, which match with the eight components of the state $b^i_{-\frac{1}{2}}|0,k\rangle$. To demonstrate that there is an equal number of bosons and fermions at each mass level requires the following "aequatio identica satis abstrusa"

$$\frac{1}{2q} \left[\prod_{n=1}^{\infty} (1+q^{2n-1})^8 - \prod_{n=1}^{\infty} (1-q^{2n-1})^8 \right] = 8 \prod_{n=1}^{\infty} (1+q^{2n})^8 , \tag{4.9}$$

which was already known to the German mathematician Jacobi in 1829! The above relation provides an example for the connection between string theories and rather deep mathematical results.

It may seem somewhat awkward to describe the superstring by a truncation via (4.7) and (4.8), and to have to rely on identities such as (4.9) to make the supersymmetry of the spectrum explicit. It was this circumstance, and the desire to explore superstring theory further, that prompted Green and Schwarz in 1981 to develop the so-called "new formalism". Instead of gluing together the Neveu-Schwarz and the Ramond sector, and projecting out half of the states in both sectors, they replaced the oscillators b^i_r and d^i_n by a single set of integer moded anticommuting oscillators S^a_m, which carry a spinor index a=1,...,8 rather than a vector index i, and obey the anticommutation relations[16]

$$\{S^a_m, S^b_n\} = \delta^{ab}\,\delta_{m+n,0} \;. \tag{4.10}$$

Because S^a belongs to a spinor representation 8_S of SO(8), S^a_m transforms as a space-time spinor under transverse Lorentz rotations, and there are no apparent "paradoxes" any more with the space-time statistics of the states. A drawback of this new formalism is that it works so far only in the light cone gauge. The question of how to extend superstring theory "off-shell" is presently under investigation by many groups.

It is quite straightforward to work out the spectrum of the superstring by means of the new formalism. Remembering that there is no tachyon any more, we can write the massless ground state of the open superstring as

$$|i\rangle \in 8_V \quad \text{(vector representation of SO(8))} \;,$$
$$|i\rangle \in 8_S \quad \text{(spinor representation of SO(8))} \;, \tag{4.11}$$

where both indices i and a assume the values 1,....,8 and 8_V and 8_S are the usual designations for the eight-dimensional vector and spinor representations of SO(8). One could alternatively assign the spinor to the conjugate spinor representation of SO(8), which is denoted by 8_C or $|\dot{a}\rangle$. (The group SO(8) is unique in that it has three inequivalent eight-dimensional representations. This property is referred to as "triality" (see e.g. Slansky[30]).) The excited states are obtained by acting on either $|i\rangle$ or $|a\rangle$ with either α^i_{-m} or S^b_{-m}. The supersymmetry is now manifest; for example, the first excited level contains the states

$$\text{128 Bosons:} \quad \alpha^i_{-1}|j\rangle \text{ and } S^a_{-1}|b\rangle \;,$$
$$\text{128 Fermions:} \quad \alpha^i_{-1}|b\rangle \text{ and } S^a_{-1}|j\rangle \;. \tag{4.12}$$

One of the accomplishments of Green and Schwarz[16] was the demonstration that one could construct the generators of the full Lorentz group SO(1,9) in ten dimensions out of just the transverse oscillators α^i_m, S^a_m and the center-of-mass coordinates and momenta.

Closed superstrings can be constructed in complete analogy with the closed bosonic string discussed in section 3. One simply has to double the number of oscillators and ground states, so we have bosonic oscillators α^i_m, $\bar{\alpha}^i_m$. The ground states are now obtained by decomposing the products

$$\begin{aligned} &\text{Bosons} \quad |i\rangle_L \otimes |j\rangle_R \quad \text{and} \quad |a\rangle_L \otimes |b\rangle_R \\ &\text{Fermions} \quad |i\rangle_L \otimes |b\rangle_R \quad \text{and} \quad |a\rangle_L \otimes |j\rangle_R \end{aligned} \tag{4.13}$$

into irreducible components. In particular, (4.13) contains the following states

$$\begin{aligned} &\text{"Graviton"} = \text{symmetric traceless part of } |i\rangle_L \otimes |j\rangle_R \\ &\text{"Gravitino"} = \gamma\text{-traceless parts of } |i\rangle_L \otimes |b\rangle_R \text{ and } |a\rangle_L \otimes |j\rangle_R \end{aligned} \tag{4.14}$$

Thus we have two gravitinos, and since the massless states form a supermultiplet, it is no surprise that the full set of states (4.13) coincides with an N=2 supergravity multiplet in ten dimensions. Hence, in the same way as there is no closed string theory without gravity, there is no closed superstring theory without supergravity!.

We have already mentioned that one may alternatively assign the ground state spinor to the conjugate representation 8_C of SO(8). In (4.13), both the left- and right-moving spinors belong to the 8_S representation, and the resulting theory is called "type II B". To get the so-called "type II A" theory, one must assign these spinors to different representations, and in this case the ground states are obtained from the products

$$\begin{aligned} &\text{Bosons} \quad |i\rangle_L \otimes |j\rangle_R \quad \text{and} \quad |a\rangle_L \otimes |\dot{b}\rangle_R \\ &\text{Fermions} \quad |i\rangle_L \otimes |\dot{b}\rangle_R \quad \text{and} \quad |a\rangle_L \otimes |j\rangle_R \end{aligned} \tag{4.13'}$$

Again, one gets an N=2 supergravity multiplet in this way.

Both the "type II A" and the "type II B" superstring theories are one-loop finite and free of anomalies, and therefore good candidates for a unified theory. However, there is another kind of string theories with these properties, namely the heterotic string theories (Gross et al.[21]). These are hybrids of the bosonic string in 26 dimensions and the superstring in ten dimensions. The most general solution to the free wave equation (2.10) is a superposition of left-moving and right-moving modes, which only depend on $\sigma-\tau$ and $\sigma+\tau$, respectively, as is also evident from the closed string mode expansion (2.12). Since (2.10) contains no interactions, these do not interfere with each other, and may therefore be chosen independently. The basic idea is now to take the left-moving sector to be a superstring with states $(8_V)_L$ and $(8_S)_L$, and the right moving sector to be a bosonic string, and to obtain the states of the full theory by a multiplication analogous to (4.13) and (4.14), but now with one half of the string in ten dimensions and the other in 26 dimensions. Absurd as it may appear at first sight, this idea does work! The crucial ingredient that makes it work is a "compactification" of the 26-dimensional part, by which the momentum components k^I with $11 \leq I \leq 26$ become discrete. They are then no longer interpreted as momenta, but rather as internal symmetry labels, such as isospin and strangeness quantum numbers. It is in this way that an internal symmetry is generated out of a purely bosonic theory which contains no internal symmetry (Frenkel and Kac[31]; Goddard and Olive[32]). Consistency (i.e. (for experts) modular invariance) then forces these symmetry groups to be either SO(32) or $E_8 \times E_8$, in accordance with the previously found restrictions from anomaly cancellations.

5. INTERACTING STRINGS

So far we have described free string theories. Knowing the spectrum, one would also like to calculate scattering amplitudes and other quantities of interest. To do so, one must develop a formalism for interacting strings. In this section, we will very sketchily explain how this can be done, mostly by drawing pictures. It is a rather demanding task to translate these pictorial representations of interacting strings into some kind of calculational scheme, and any attempt at a more detailed explanation would go far beyond the limitations of this article. The interested reader is referred to Mandelstam[33,34] and Cremmer and Gervais[35] for further details of the formalism. The essential result is that string interactions are very restricted and almost unique.

To understand the basic idea, it is useful to go back once more to the relativistic point particle which was discussed in section 2. Its interactions can be very simply represented as splitting of world-lines, as in Fig. 6 below.

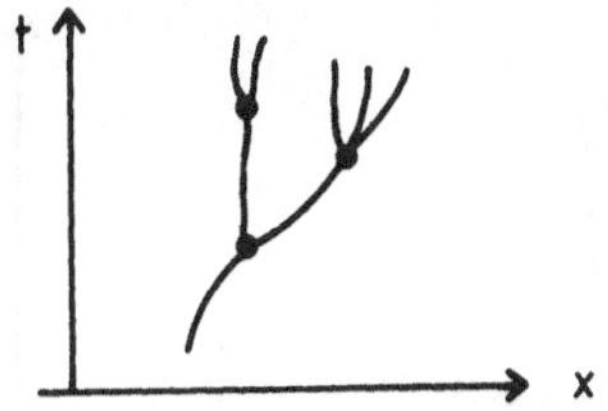

Fig.6. Interactions of the relativistic point particle

It is very important that, for the relativistic point particle, there are essentially no limitations on such interactions: the world line may split at any of its points and branch off into arbitrarily many new world-lines. The mathematical description of such an interaction is through a vertex

$$V(x_0^\mu(\tau_0),x_1^\mu(\tau_0),\ldots,x_n^\mu(\tau_0)) = \delta(x_0^\mu(\tau_0)-x_1^\mu(\tau_0))\ldots\delta(x_0^\mu(\tau_0)-x_n^\mu(\tau_0)), \quad (5.1)$$

where the world line of an incoming particle (parametrized by $x_0^\mu(\tau)$) splits up into the world lines of n particles (parametrized by $x_1^\mu(\tau),\ldots,x_n^\mu(\tau)$) at the interaction time $\tau=\tau_0$. A "radiative correction" is obtained by splitting a world-line and joining the pieces at a later time, see Fig. 7. In fact, these pictures are nothing but ordinary Feynman diagrams, and the knowledgable reader will recall that it is not completely straightforward to translate these pictures into mathematically well-defined expressions.

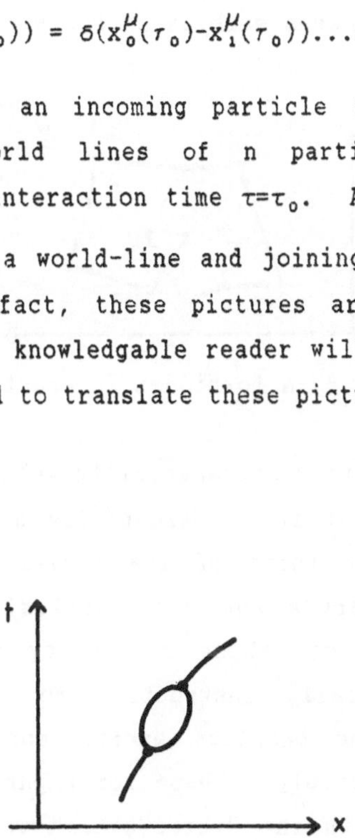

Fig.7. A one-loop correction for the relativistic point particle

At this level of the discussion, the interaction of the strings are quite similar to the point particle interactions. Strings interact by touching at one point and joining to a single string: for the open string the interaction point is always the boundary point, while for the closed string the point of contact is arbitrary. These processes are depicted in Fig. 8. below.

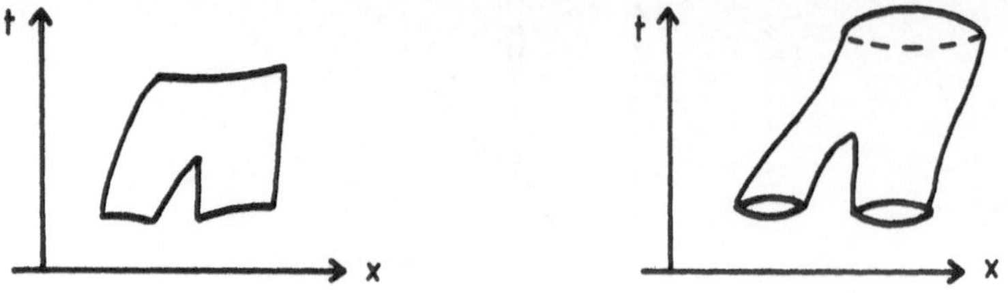

Fig.8. Interactions of open and closed strings

A rather important point here is that although these are the interactions between extended objects, the interaction itself is local: the instantaneous interaction takes place at one point only. The locality postulate rules out processes such as in Fig.9.

Fig.9. A forbidden interaction

One can now associate a mathematically well-defined "vertex-operator" with such an interaction; it is essentially a string overlap δ-function analogous to (5.1). In terms of the individual states of the string theory (parts of which are shown in the tables), one gets infinitely many point particle interactions whose complexity increases with increasing "spin". It is especially instructive to calculate the three-point interactions between the massless excitations of the open and closed bosonic strings, respectively. These point particle interaction vertices turn out to coincide with the "three-gluon" vertex of Yang-Mills theory (Neveu and Scherk[36]) in the case of the open string, and with the "three graviton" vertex of Einstein's general relativity theory in the case of the closed string (Scherk and Schwarz[7]). This means that

(i) Ordinary Yang Mills theories are contained in the open bosonic string theory, and

(ii) Einstein's relativity theory is contained in the closed bosonic string theory.

In a sense, one could have foreseen this result: the only consistent theories of massless particles of spin-1 and spin-2, respectively, are Yang-Mills theories and Einstein's general relativity, respectively (the gauge invariance is absolutely necessary to eliminate unwanted helicity states). Nonetheless, the reader should pause at this point to appreciate the implications of this result. After all, the massless states are only a tiny part of the whole string spectrum, and one may therefore anticipate the existence of a much bigger symmetry which contains either ordinary gauge symmetries or general coordinate invariance as "the tip of the iceberg". It is one of the most fascinating problems of string theory what this huge symmetry might be, and whether there is a generalization of the principle of equivalence that encompasses the postulates of general relativity.

Similar remarks apply to superstrings, whose interactions are also given by overlap δ-functions (Mandelstam[34]; Green and Schwarz[37]). However, these are now harder to visualize, and we will therefore refrain from drawing further diagrams. As before, one may calculate the point-like interactions between massless particles. The result is that

(iii) Ordinary supersymmetric Yang Mills theories are contained in the open superstrings, and

(iv) Supergravity is contained in the closed superstring theory.

It is now obvious why superstring theory has completely absorbed the once thriving field of supergravity (to be sure, there is one supergravity theory that does not fit into string theory, namely the maximally extended d=11 supergravity[38]).

Finally, radiative corrections can be discussed along similar lines. They correspond to first splitting and rejoining strings, see Fig.10.

Fig.10. A one-loop correction for the closed bosonic string

The number of loops is equal to the number of holes in the world-surface of the string. Possible divergences appear when the diameter of such a hole shrinks to zero, or when the surface is "pinched"; this is somewhat analogous to the divergences that appear in Feynman graphs when a loop shrinks to a point. It is conjectured, although not proven so far, that, in contrast to point particle theories, the one-loop finiteness of string theories implies their finiteness to *all* orders of pertubation theory.

6. OUTLOOK

Up to now, we have concentrated on the basic features of string theory, namely those that would be included in any introductory treatment of the subject. However, there are many more advanced topics and, of course, many open problems. In this section we shall try to give the reader a flavor of what these are, but naturally our review will be incomplete. The areas of current research can be roughly divided into two parts. The first consists of attempts to extract physically testable predictions from superstring theory, while the second centers on the underlying principles of string theory. Let us begin with the first.

As already mentioned, the theory currently thought to be most promising is the heterotic string with gauge group $E_8 \times E_8$[21]. Although there exist other versions of the heterotic string, this theory is particularly attractive for phenomenology. The group E_8 is the largest of the exceptional Lie-groups, and is big enough to accomodate all known symmetries and particles; for this reason, it has already been considered for grand unification several years ago. Furthermore, the $E_8 \times E_8'$ theory has chiral fermions and is free of anomalies (this is also true for the other heterotic theories as well as for the type IIB theory). Thus it offers the possibility of getting chiral fermions in four dimensions in the desired representations of SU(3)xSU(2)xSU(1). The way this is achieved in practice is related to the way in which the ten-dimensional theory is compactified to four dimensions. In the process of compactification, six dimensions are curled up to an "internal" manifold whose size is so small as to make it inaccessible to present day experiments (e.g. with diameter of the order of 10^{-33} cm). The number of chiral fermions which emerge in such a compactification is related to topological properties of the internal manifold, i.e. the number of its "holes" and "handles". This is an example of how qualitative features of our low energy world, such as the number of generations, may be linked to topological rather than metrical properties of a unified theory.

In a currently favoured scenario[22] the compactification occurs on a "Calabi-Yau manifold" (these manifolds are mathematically rather intricate and interesting objects, but there is no room here to discuss them in further detail), and the gauge group $E_8 \times E_8'$ is broken according to

$$E_8 \times E_8' \rightarrow G \times E_8' , \tag{6.1}$$

where the residual gauge group G is a subgroup of E_6. All observed particles (quarks, leptons,etc.) transform under G, whereas the particles associated with E_8' are almost completely decoupled from our known universe, as they couple only gravitationally. The E_8' particles constitute a "shadow world" (thus we may be sitting in the middle of a "shadow mountain" without noticing it!). This is interesting, because invisible "shadow-matter" may account for the dark matter whose origin is still not understood by astrophysicists. The observable group G must still be further broken to the standard low energy group $SU(3)_C \times SU(2) \times SU(1)_Y$, and it is here that things get murky. Since the actual dynamics of the theory is unknown, one has to make many assumptions at this point, and the outcome of any calculation depends to a great extent on the assumptions that were put in at the beginning. A second problem is that the compactification on Calabi-Yau spaces is not unique; the number of solutions is astronomical, and one can obtain almost any number of chiral generations depending on the topology of the Calabi-Yau manifold. One would rather prefer to have a unique solution to describe the compactification to our four-dimensional world. Another problem is that compactification to four dimensions is in no way preferential in superstring theories (unlike in the case of D=11 supergravity where four dimensions are preferred[39]. It seems obvious that the solution of these problems will require a lot more work.

We next turn to the second area of problems having to do with questions of principle. String theories possess many "miraculous" properties, which were usually found after long and arduous calculations, especially in the early days of the subject. For instance, why does the massless state in the closed string theory behave like a graviton? We know that the only consistent theory of a massless spin-2 field is Einstein's theory of general relativity, so even if we start with a massless free spin-2 field, and try to make it interact, we must eventually bring in the full apparatus of Riemannian geometry. Of course, Riemannian geometry and the principle of equivalence were Einstein's points of departure, and it was only realized afterwards that the graviton was a massless spin-2 particle. However, in string theories we lack both the analog of Riemannian geometry and a generalized principle of equivalence, and so we must start from the other end. An indication that this may be the "wrong" end to start from is that until now we are only able to describe the string motion in a fixed (not necessarily flat) spacetime background, while the string itself contains the seeds of curved spacetimes with nontrivial topology, and should therefore be described in a much more general way. It seems therefore that in order to properly describe strings one must dissolve the very notion of spacetime in the same way that quantum mechanics does away with the notion of the trajectory of an electron in the hydrogen atom. These conceptual problems are presently at the focus of research, but it is not clear how long it will take to solve them.

One attempt in this direction is covariant string field theory (see West[40] for a recent review). The purpose here is to exhibit the invariances explicitly which generalize ordinary gauge invariance and invariance under general coordinate transformations. To illustrate the basic idea we introduce a "string functional", which associates a field with every string action according to

$$\Psi = [\ \varphi(x) + A_\mu(x)\alpha^\mu_{-1}\ ]\ \dots\ ]|0\rangle\ . \tag{6.2}$$

The physical state constraint $L_1\Psi = 0$ can be easily evaluated using $L_1 = p_\nu\alpha^\nu_1 + \dots$ and the basic commutator (3.1),

$$0 = L_1\Psi = (p_\nu\alpha^\nu_1 + \dots)(\varphi + A_\mu\alpha^\mu_{-1} + \dots)|0\rangle = (p^\mu A_\mu + \dots)|0\rangle. \tag{6.3}$$

Hence, (6.3) implies the Landau gauge condition

$$\partial^\mu A_\mu(x) = 0\ . \tag{6.4}$$

We can release this gauge condition by introducing a gauge invariance associated with L_{-1}. To do so, we define a "gauge parameter string functional"

$$\Omega = [w(x) + w_\mu(x)\alpha^\mu_{-1} + \dots\ ]|0\rangle\ . \tag{6.5}$$

Using $L_{-1} = p_\nu \alpha^\nu_{-1} + \ldots$, we see that

$$\delta\Psi = (\delta\varphi + \delta A_\mu \alpha^\mu_{-1} + \ldots)|0\rangle = L_{-1}\Omega$$

$$= (p_\nu \alpha^\nu_{-1} + \ldots)(w + w_\mu \alpha^\mu_{-1} + \ldots)|0\rangle = (p_\mu w \alpha^\mu_{-1} + \ldots)|0\rangle \tag{6.6}$$

contains the transformation rule

$$\delta A_\mu = i\partial_\mu w \ , \tag{6.7}$$

which is just the ordinary gauge transformation of the electromagnetic potential! From (6.6), we also see that $L_{-1}\Omega$ contains further transformations for the higher level fields in the expansion (6.2), and therefore an infinite tower of symmetries. But there is even more symmetry, because there will be similar transformations for all L_{-m} with $m \geq 1$. This explicitly shows the "explosion of symmetry" in string theories which was alluded to in the Introduction.

The main task is now to work out the fully gauge invariant action, i.e. the string analog of $F_{\mu\nu}F^{\mu\nu}$ with $F_{\mu\nu} = \partial_\mu A_\nu - \partial_\nu A_\mu$, first at the free level, and then for the interacting theory (which should in particular contain the three- and four-gluon vertices), and to repeat this exercise for superstring theory. A great deal of progress has been made during the last year, although it is probably too early to tell whether the conceptual breakthrough can be achieved in this way. However, apart from such considerations, one may anticipate that the formalism will be eventually useful in studying higher loop corrections, and in finding classical and/or nonperturbative solutions to string theory.

A further question of considerable interest is why there are already so many string theories (about ten at the time of writing) that are fully consistent at the one-loop level (and presumably beyond), where one would be enough, and whether these theories are related. In Freund[41], Casher et al.[42], and in Englert, Nicolai and Schellekens[43], it has been suggested that all consistent superstring theories are just spontaneously broken versions of the purely bosonic D=26 string theory, which should therefore be viewed as the "Ur-theory". In fact, it has been established there that superstrings are contained in the bosonic string, but the question as to the dynamical origin of this symmetry breaking remains open. Again, much work is needed to make progress.

Finally, we should not close our eyes to the possibility that the final string theory may not yet have been found, or that there exists a theory "beyond superstrings". While efforts in this direction have not borne fruit so far, one may safely predict that the coming years will have some surprises in store, which may change the course of theoretical high energy physics and our perception of it in unexpected ways.

REFERENCES

Early Review Articles

V. D. Allessandrini, D. Amati, M. le Bellac and D. Olive, Phys. Rep. 1C: 269 (1971).

S. Mandelstam, Phys. Rep. 13C: 259 (1974).

C. Rebbi, Phys. Rep. 12C: 1 (1974).

J. Scherk, Rev. Mod. Phys. 47: 123 (1975).

J. H. Schwarz, Phys. Rep. 8C: 269 (1973).

References Listed in Text

1. G. Veneziano, Nuovo Cim. 57A: 190 (1968).
2. Y. Nambu, in "Proceedings Int. Conf. on Symmetries and Quark Models", Gordon and Breach, (1970).
3. L. Susskind, Nuovo Cim. 69A: 457 (1970).
4. H. Fritzsch, M. Gell-Mann and H. Leutwyler, Phys. Lett. 47B: 365 (1973).
5. R. C. Brower, Phys. Rev. D6: 1655 (1972).
6. P. Goddard and C. B. Thorn, Phys. Lett. 40B: 2; 235 (1972).
7. J. Scherk and J. H. Schwarz, Nucl. Phys. B81: 118 (1974).
8. J. Wess and B. Zumino, Nucl. Phys. B70: 39; Phys. Lett. 49B: 52 (1974).

9. S. Ferrara, D. Freedman and P. van Niuwenhuizen, Phys. Rev. D13: 3214 (1976).

10. S. Deser and B. Zumino, Phys. Lett. 62B: 335 (1976).

11. E. Cremmer and B. Julia, Nucl. Phys. B159: 141 (1979).

12. B. de Wit and H. Nicolai, Nucl. Phys. B208: 322 (1982).

13. P. Ramond, Phys. Rev. D3: 2415 (1971).

14. A. Neveu and J. H. Schwarz, Nucl. Phys. B31: 86 (1971).

15. F. Gliozzi, J. Scherk and D. Olive, Nucl. Phys. B122: 253 (1976).

16. M. B. Green and J. H. Schwarz, Nucl Phys. B181: 502 (1981).

17. M. B. Green and J. H. Schwarz, Nucl. Phys. B198: 252; 441 (1982).

18. L. Alvarez-Gaumé and E. Witten, Nucl Phys. B234: 269 (1983).

19. M. B. Green and J. H. Schwarz, Phys. Lett. 149B: 117 (1984).

20. J. Thierry-Mieg, Phys. Lett. 156B: 199 (1985).

21. D. J. Gross, J. A. Harvey, E. Martinec and R. Rohm, Nucl. Phys. B256: 253 (1985).

22. P. Candelas, G. T. Horowitz, A. Strominger and E. Witten, Nucl. Phys. B258: 46 (1985).

23. P. A. M. Dirac, Can. J. Phys. 2: 129 (1950).

24. J. Scherk, Rev. Mod. Phys. 47: 123 (1975).

25. A. M. Polyakov, Phys. Lett. 103B: 207;211 (1981).

26. P. Goddard, J. Goldstone, C. Rebbi and C.B. Thorn, Nucl. Phys. B56: 109 (1973).

27. R. Hagedorn, Nuovo Cim. 56A: 1027 (1968).

28. S. Coleman, Phys. Rev. D11: 2088 (1975).

29. S. Mandelstam, Phys. Rev. D11: 3026 (1975).

30. R. Slansky, Phys. Rep. 79C: 1 (1981).

31. J. Frenkel J. and V. Kac, Inv. Math. 62: 23 (1980).

32. P. Goddard and D. Olive, in "Vertex Operators in Mathematics and Physics", Springer Verlag 51 (1984).

33. S. Mandelstam, Nucl. Phys. B64: 205 (1973).

34. S. Mandelstam, Nucl. Phys. B69: 77 (1974).

35. E. Cremmer and J. L. Gervais, Nucl. Phys. 876: 209 (1974).

36. A. Neveu and J. Scherk, Nucl. Phys. B36: 155 (1972).

37. M. B. Green and J.H. Schwarz, Nucl. Phys. B218: 43 (1983).

38. E. Cremmer, B. Julia and J. Scherk, Phys. Lett. 76B: 409 (1978).

39. P. G. O. Freund and M. A. Rubin, Phys. Lett. 97B: 233 (1980).

40. P. C. West, "Gauge covariant String Field Theory", CERN preprint TH 4460/86 (1986).

41. P. G. O. Freund, Phys. Lett. 151B: 387 (1985).

42. A. Casher, F. Englert, H. Nicolai and A. Taormina, Phys. Lett. 162B: 121 (1985).

43. F. Englert, H. Nicolai and A. Schellekens, Nucl. Phys. B274: 315 (1986).

Recent Review Articles

J. H. Schwarz, Phys. Rep. 89C: 223 (1982).

M. B. Green, Surveys of High Energy Physics 3: 127 (1983).

INDEX